CAMBRIDGE LIBRARY COLLECTION

Books of enduring scholarly value

Life Sciences

Until the nineteenth century, the various subjects now known as the life sciences were regarded either as arcane studies which had little impact on ordinary daily life, or as a genteel hobby for the leisured classes. The increasing academic rigour and systematisation brought to the study of botany, zoology and other disciplines, and their adoption in university curricula, are reflected in the books reissued in this series.

Journal of a Tour in Iceland, in the Summer of 1809

Sir William Jackson Hooker (1785–1865) was an eminent British botanist, best known for expanding and developing the Royal Botanic Gardens at Kew into a leading centre of botanic research and conservation. At the age of nineteen he undertook an expedition to Iceland, his first outside Britain. Unfortunately, all his specimens and notes were destroyed in a fire on the return voyage (described in Volume 1), but he was able, with the help of the notes made by Sir Joseph Banks on an earlier expedition, to write this account. His work was first published privately in 1811, but a second edition was published in 1813 and is reproduced here. In 1809 England and Denmark-Norway were at war, and Iceland was a Danish dependency. Volume 2 offers Hooker's first-hand observations on the relationship between the two countries, and also includes detailed descriptions of the many volcanoes on the island.

Cambridge University Press has long been a pioneer in the reissuing of out-of-print titles from its own backlist, producing digital reprints of books that are still sought after by scholars and students but could not be reprinted economically using traditional technology. The Cambridge Library Collection extends this activity to a wider range of books which are still of importance to researchers and professionals, either for the source material they contain, or as landmarks in the history of their academic discipline.

Drawing from the world-renowned collections in the Cambridge University Library, and guided by the advice of experts in each subject area, Cambridge University Press is using state-of-the-art scanning machines in its own Printing House to capture the content of each book selected for inclusion. The files are processed to give a consistently clear, crisp image, and the books finished to the high quality standard for which the Press is recognised around the world. The latest print-on-demand technology ensures that the books will remain available indefinitely, and that orders for single or multiple copies can quickly be supplied.

The Cambridge Library Collection will bring back to life books of enduring scholarly value (including out-of-copyright works originally issued by other publishers) across a wide range of disciplines in the humanities and social sciences and in science and technology.

Journal of a Tour in Iceland, in the Summer of 1809

VOLUME 2

WILLIAM JACKSON HOOKER

CAMBRIDGE UNIVERSITY PRESS

Cambridge, New York, Melbourne, Madrid, Cape Town,
Singapore, São Paolo, Delhi, Tokyo, Mexico City

Published in the United States of America by Cambridge University Press, New York

www.cambridge.org
Information on this title: www.cambridge.org/9781108030496

This edition first published 1813
This digitally printed version 2011

ISBN 978-1-108-03049-6 Paperback

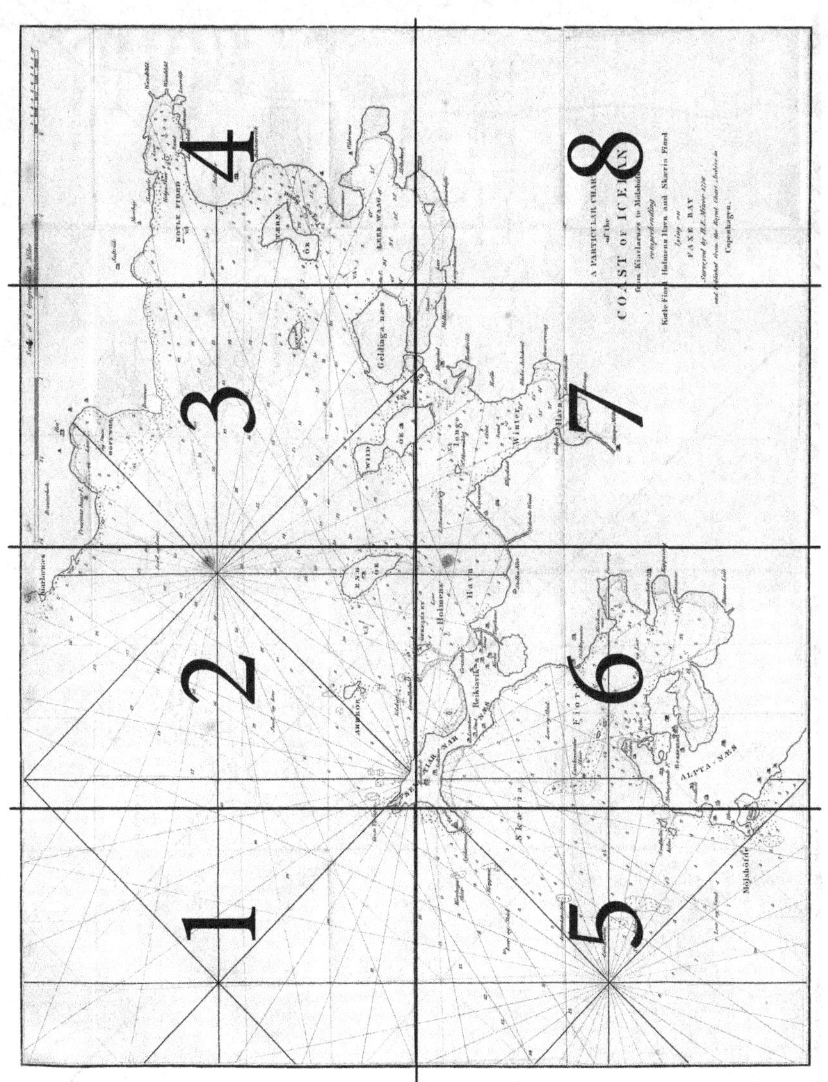

Key to following pages.

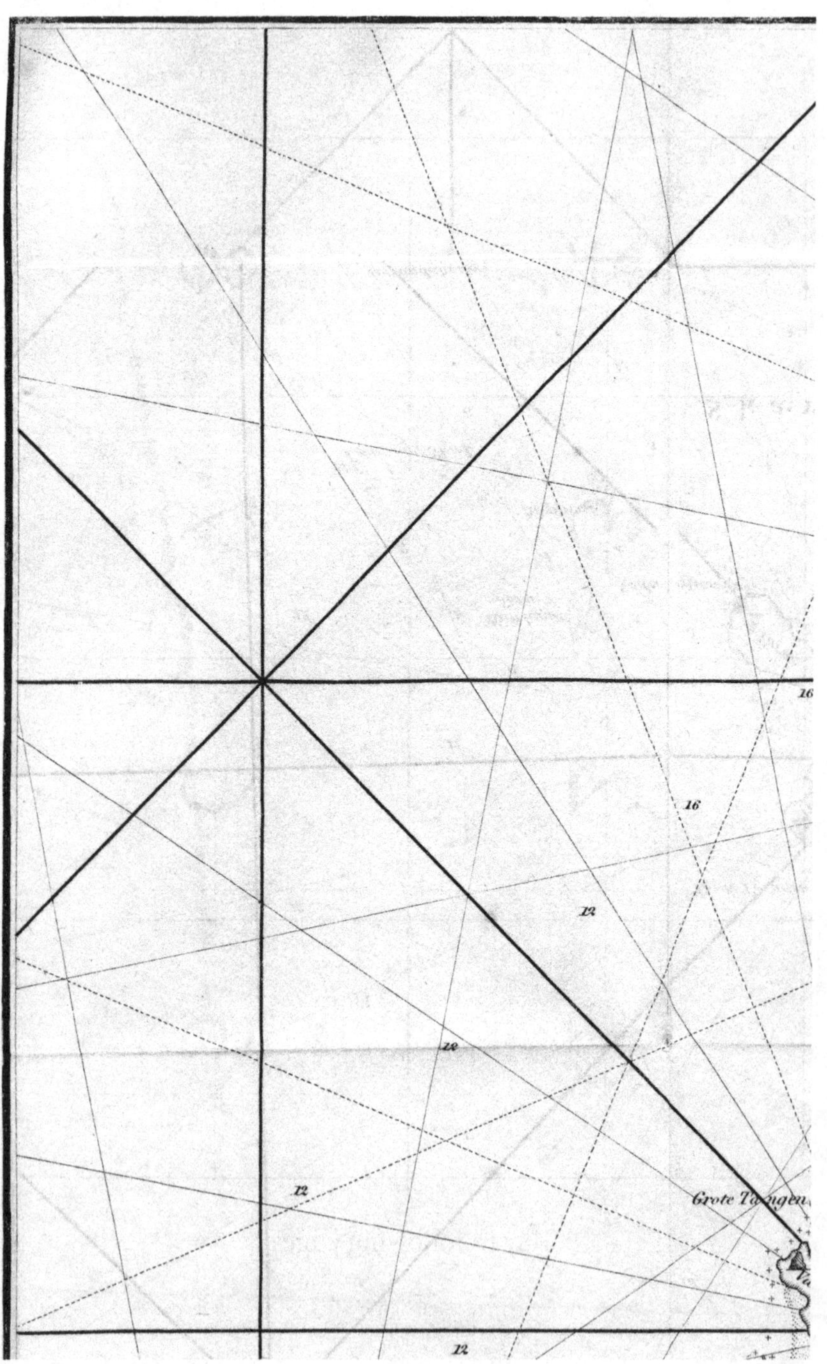
16
16
12
12
12
Grote
12

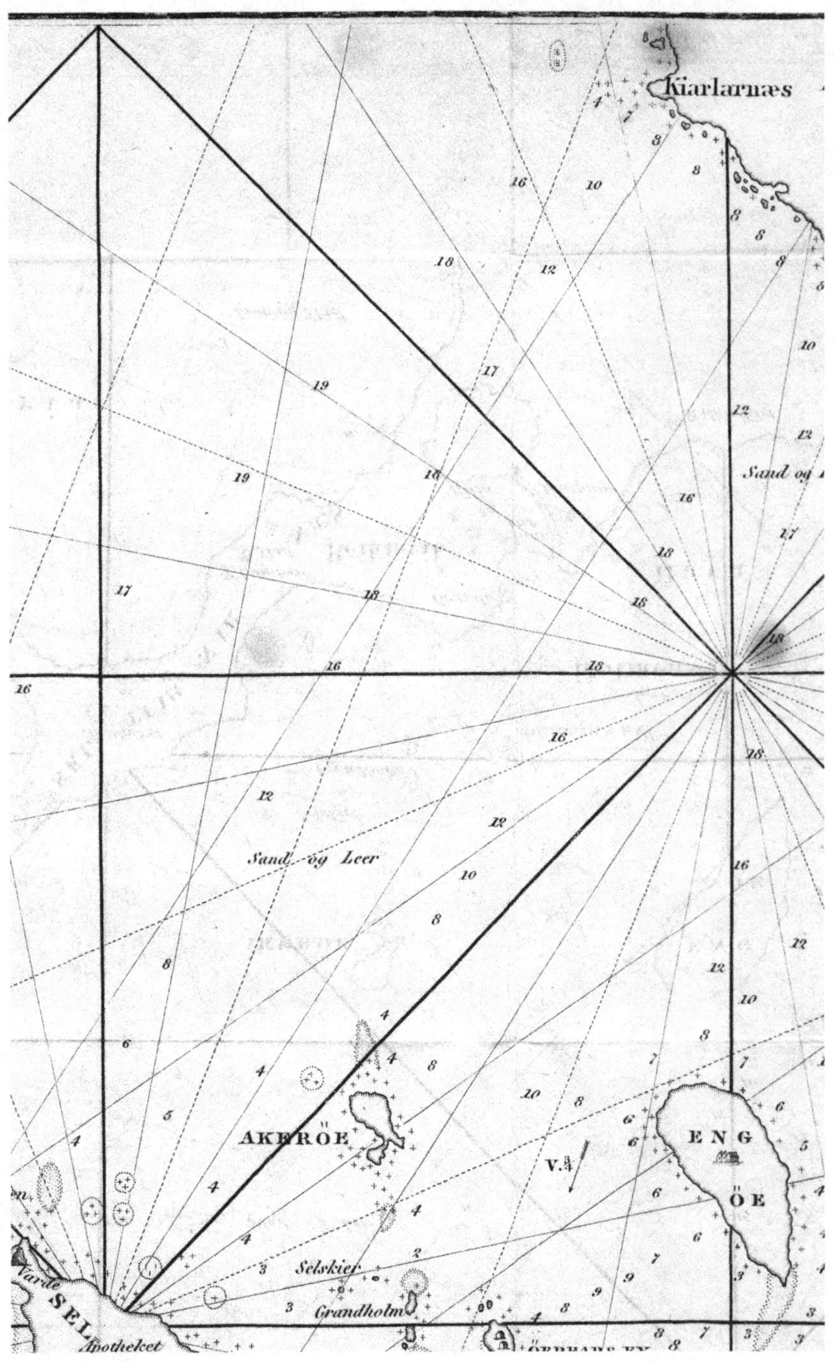
Kiarlarnæs
Sand og Leer
AKERÖE
ENG ÖE
V.¾
Selskier
Grandholm
Varde
SEL
Apotheket

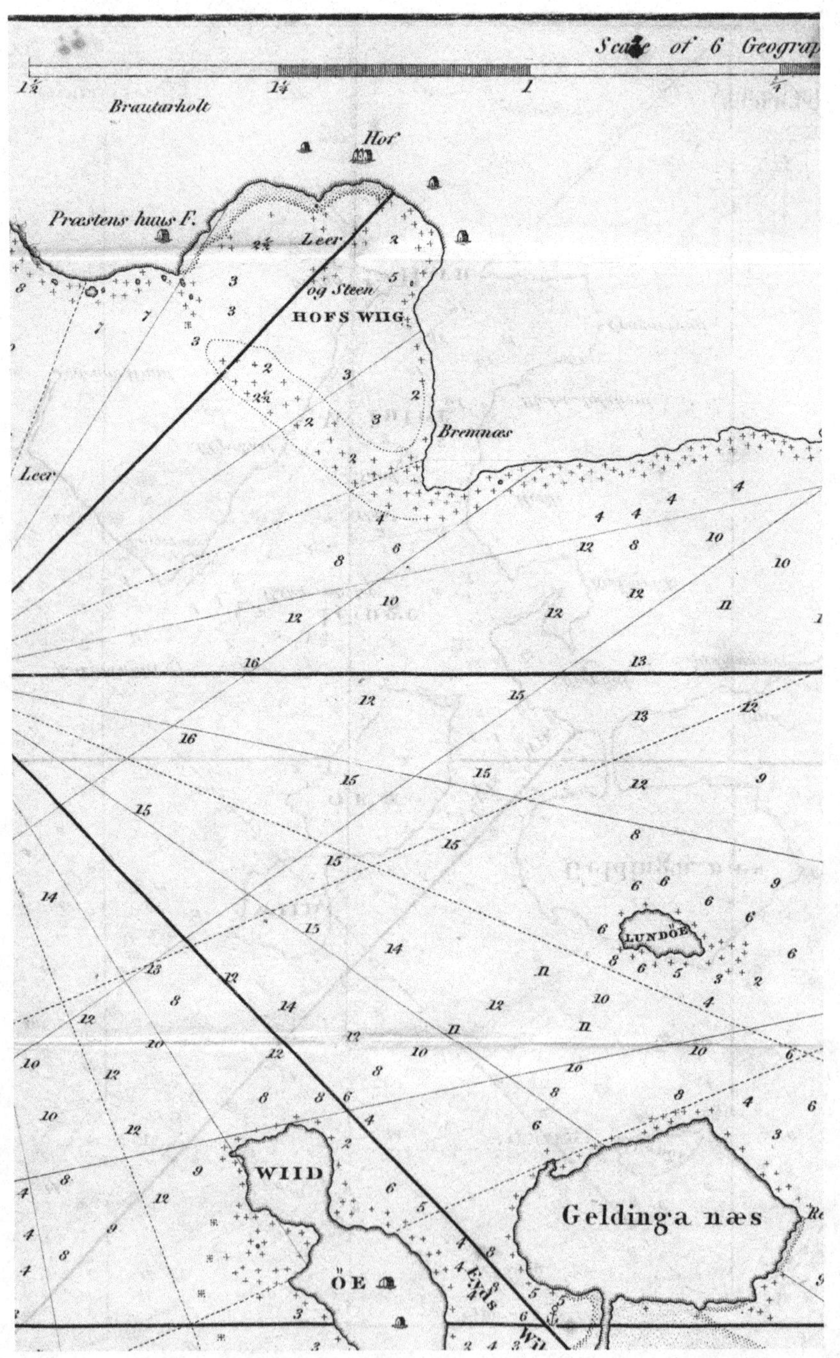

Scale of 6 Geograp
1½
1¼
1
¼
Brautarholt
Hof
Præstens huus F.
Leer
og Steen
HOFS WIIG
Bremnæs
Leer
LUNDÖE
WIID
ÖE
Geldinga næs
Eyds

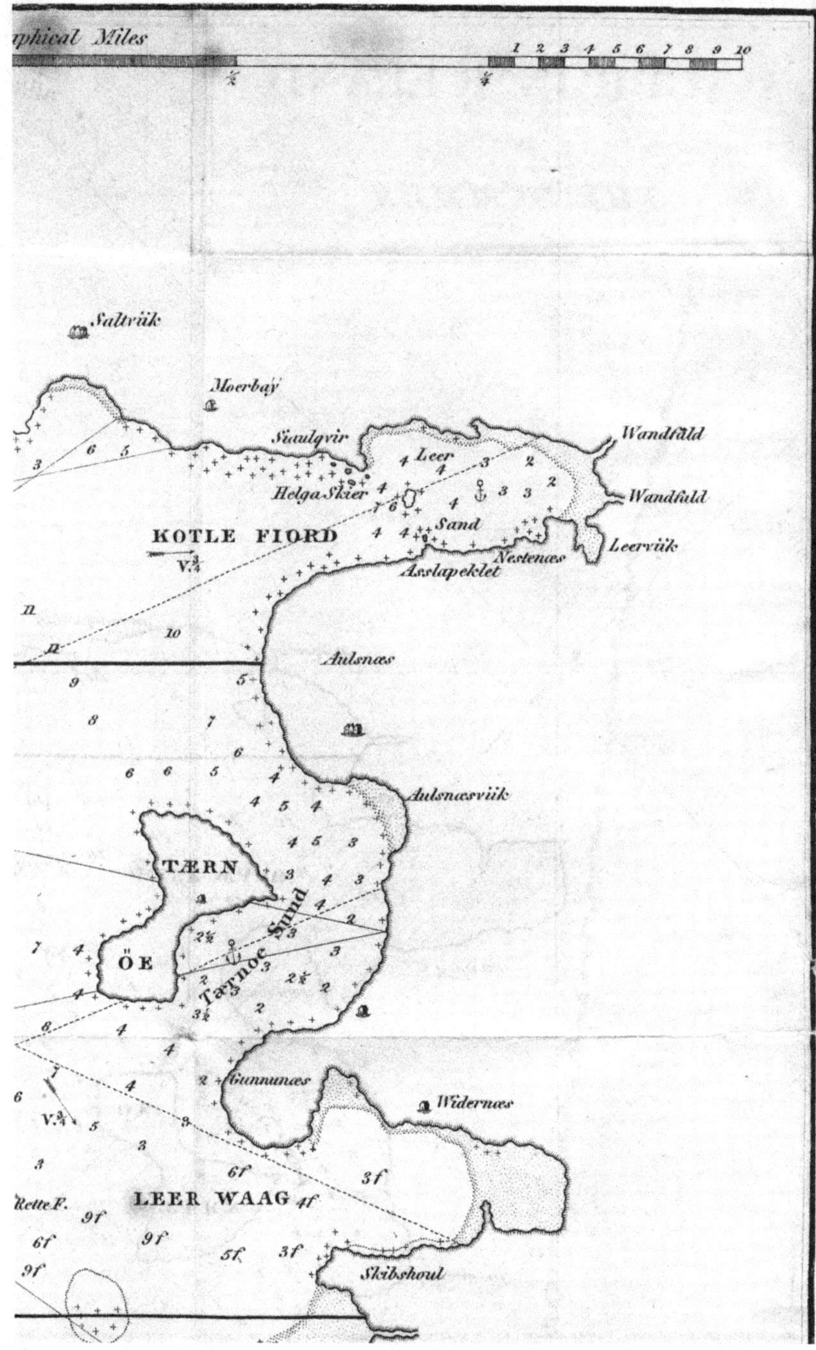

phical Miles
Saltviik
Moerbay
Siaulqvir
Leer
Wandfald
Helga Skier
Wandfald
Sand
KOTLE FIORD
Leerviik
Nestenæs
Asslapeklet
Aulsnæs
Aulsnæsviik
TÆRN
ÖE
Tarnoe Sund
Gunnunæs
Widernæs
LEER WAAG
Rette F.
Skibshoul

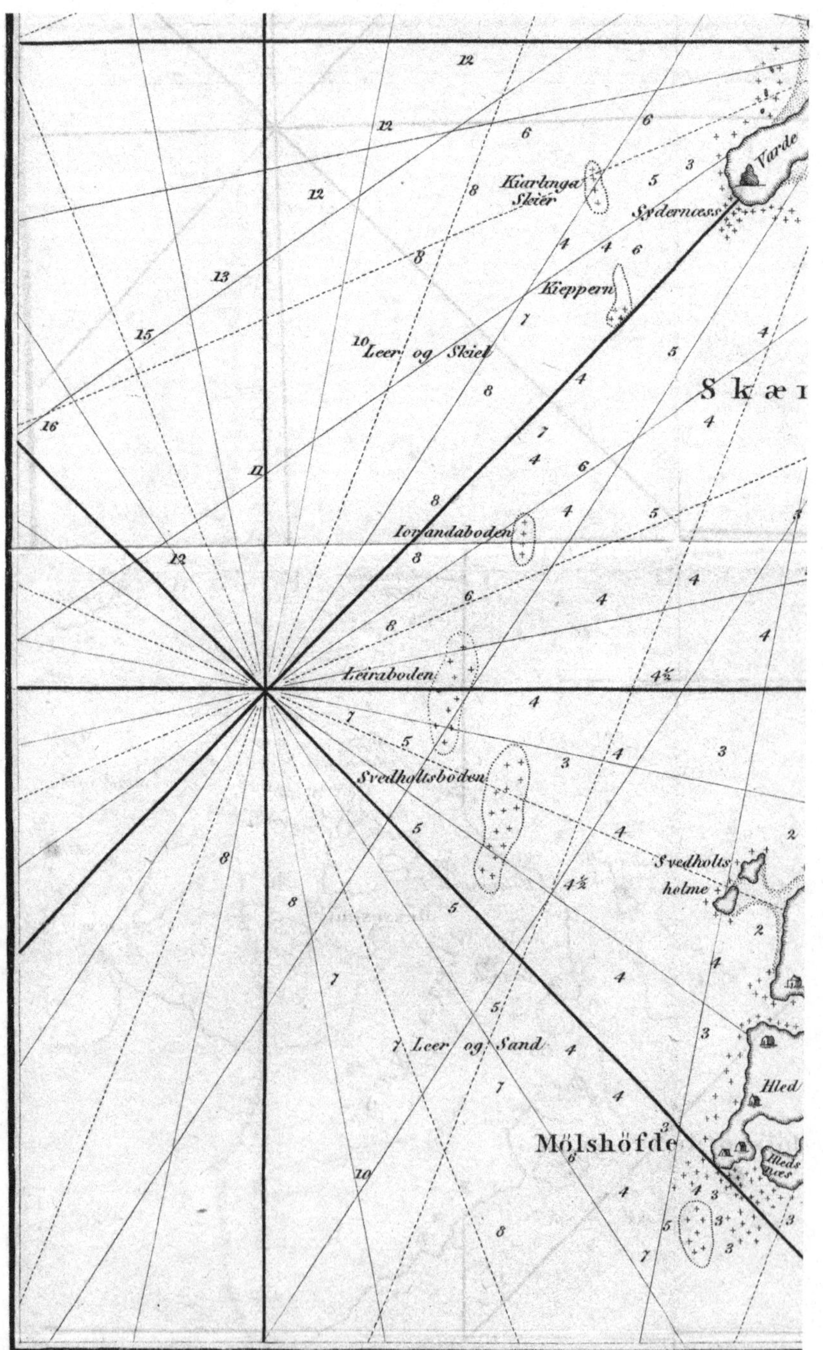
Varde
Kiarlinga
Skiër
Syderness
Kieppern
Leer og Skiel
Skæ
Iorandaboden
Leiraboden
Svedholtsboden
Svedholts
helme
Leer og Sand
Mölshöfde
Hled
Hleds
naes

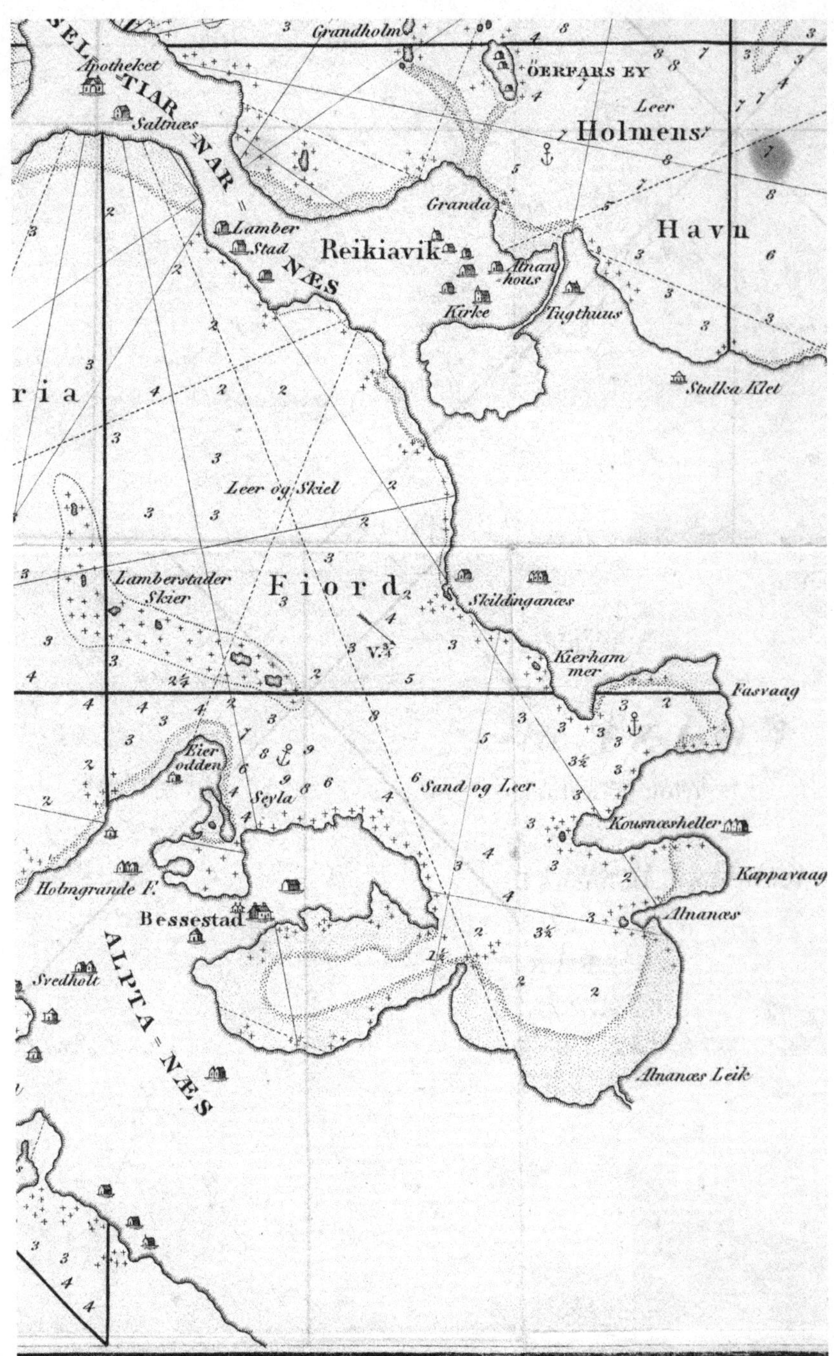
Grandholm
Apotheket
SELTIAR=NAR=NÆS
Saltnæs
ÖERFARS EY
Leer
Holmens
Havn
Granda
Lamber
Stad
Reikiavik
Alnan hous
Kirke
Tugthuus
Stulka Klet
ria
Leer og Skiel
Fiord
Lamberstader Skier
Skildinganæs
Kierhammer
Fasvaag
Eier odden
Seyla
Sand og Leer
Kousnæsheller
Kappavaag
Holmgrande F.
Bessestad
Alnanæs
Svedholt
ALPTA=NÆS
Alnanæs Leik

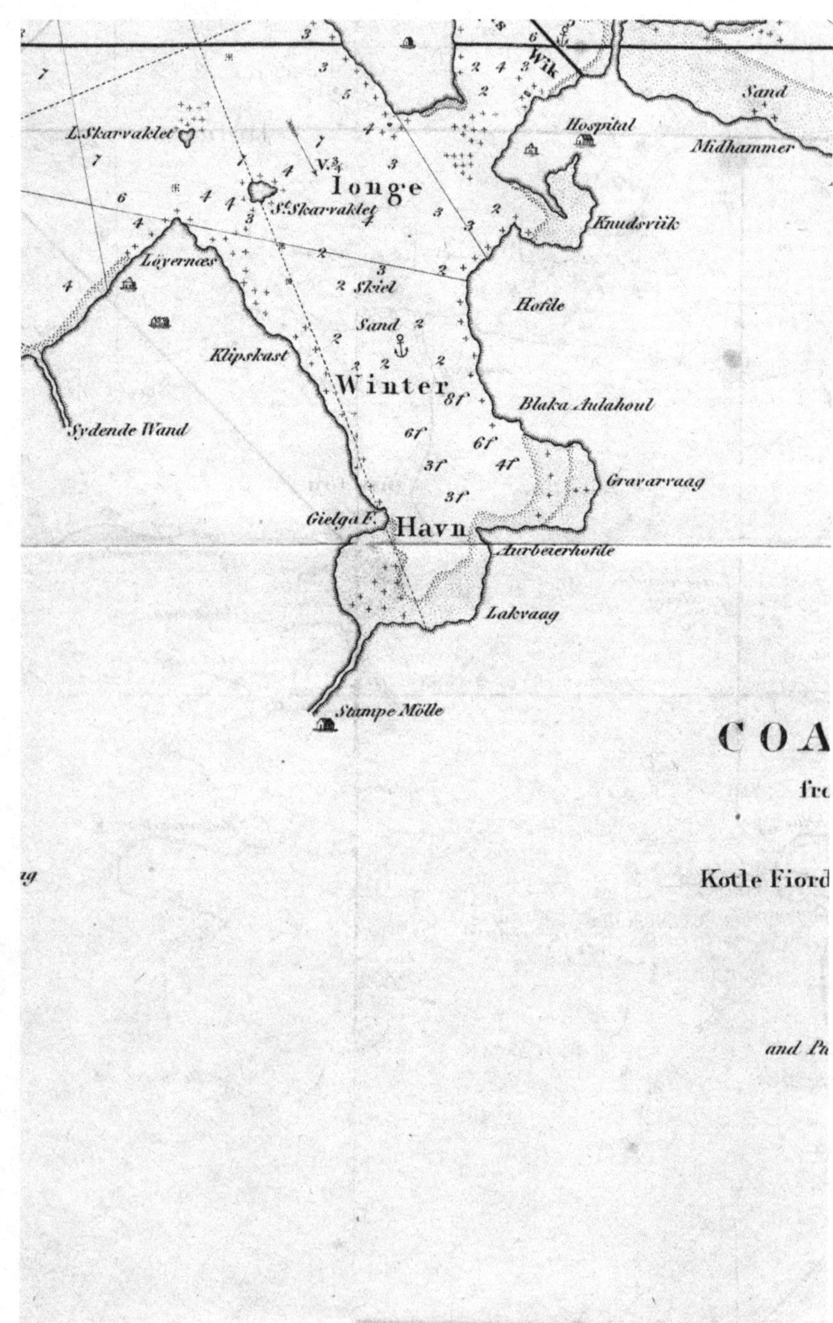
L.Skarvaklet
Ionge
St. Skarvaklet
Löyernæs
Klipskast
Sydende Wand
Skiel
Sand
Winter
Havn
Gielga F.
Stampe Mölle
Hospital
Midhammer
Sand
Wik
Knudsviik
Hofile
Blaka Aulahoul
Gravarraag
Aurbeierhofile
Lakvaag
COA
fr
Kotle Fiord
and P

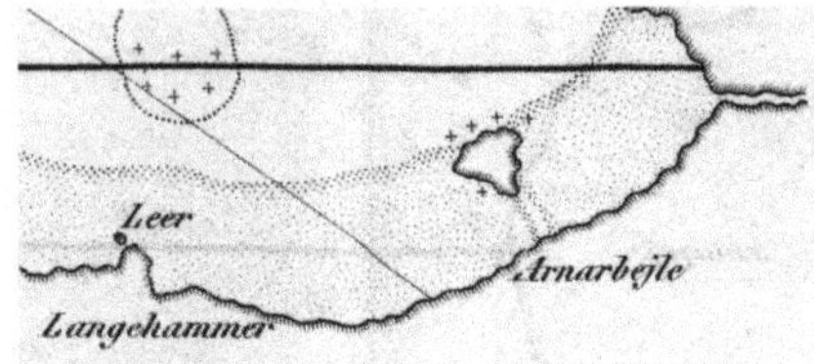

A PARTICULAR CHART

of the

ÀST OF ICELAND

·om Kiarlarnæs to Mölshöfde

comprehending

·d Holmens Havn and Skæria Fiord

lying on

FAXE BAY

Surveyed by H.E. Minor 1776

Published from the Royal Chart Archive in

Copenhagen.

JOURNAL

OF

A TOUR IN ICELAND,

IN

THE SUMMER

OF

1809.

BY

WILLIAM JACKSON HOOKER, F. L. S.

AND

FELLOW OF THE WERNERIAN SOCIETY OF EDINBURGH.

SECOND EDITION, WITH ADDITIONS.

VOL. II.

LONDON:

PRINTED FOR LONGMAN, HURST, REES, ORME, AND BROWN, PATERNOSTER-ROW,

AND JOHN MURRAY, ALBEMARLE-STREET,

By J. Keymer, Yarmouth.

1813.

APPENDIX. A.

DETAIL

OF THE

ICELANDIC REVOLUTION

IN 1809.

RECOLLECTIONS

OF

ICELAND.

APPENDIX. A.

TRIFLING and insignificant as every thing connected with the politics of so small and so miserable a country as Iceland must necessarily appear, when contrasted with the events that are agitating the great powers of Europe, nevertheless, as the government of this island underwent a total change during the short time of my residence in it, I feel, not only that my journal would be incomplete were I to pass over such things in silence, but also that it may reasonably be required of me to give an account of transactions, which fell under my own observation, and of which, as a mere by-stander, I may be expected to speak with more impartiality than those who were actively engaged in them. I shall therefore endeavor to do it as plainly and succinctly as possible, trusting that, unimportant as are the events to be detailed in my narrative, they may not on that account be wholly devoid of interest, but may find some shelter under the old adage, that "inest sua gratia parvis." From one error, at least, that is but too common to writers of all descriptions, whatever be their subject, I flatter myself I shall be allowed to have steered clear, that of magnifying occurrences, so as to exemplify the fable of the mountain in labor; for the very reverse is my case, and I ought, perhaps, rather to dread the having fallen into the opposite extreme; as my inclinations, as well as my feelings, would have led me to have confined myself principally to the leading objects of my voyage, the natural history of the island and the manners and customs of the inhabitants, could I but have persuaded myself that I could have done so with propriety. Having, as just observed, taken no part whatever in politics, and having frequently been engaged in excursions at a distance from Reikevig, I am of course ignorant of much that passed there, and it must be remembered that the portion of my narrative that rests upon my own authority is far from great; but the remainder I have been enabled to fill up in a manner at least equally authentic, having been furnished with various documents through the kindness of Captain Jones, as well as with a complete statement * of the whole by Count Tramp, drawn up with the view of being laid before the British government, and with a similar, but more

* This statement was originally accompanied by a considerable number of letters, protests, &c., to which it refers in almost every page, but which I have never seen, and I may, probably, from this cause, have been led to do less justice to the count than would have been the case, had I had an opportunity of consulting them. It is necessary at the same time to remark, that, of the events which took place after the imprisonment of the count, he only speaks from these documents, or from information which he received verbally from the inhabitants of Reikevig, a few days previous to his leaving Iceland, and this may account for some passages which appear to me to be exaggerated, and which, had the circumstances been related from the count's own knowledge, would not have crept into his narrative.

extended, statement by Mr. Jorgensen *, detailing at full length, not only the things that occurred, but the causes that preceded and gave birth to them. Thus, then, pro-

* This gentleman I have already had occasion to mentioned more than once in my journal; but, as he has, in what follows, to appear as the principal actor, it is right to give some farther account of him; that the transaction may be shewn in its proper light, and that it may not be thought that Mr. Phelps, a subject of Great Britain, has, by taking a part in a matter unauthorised by his country, transgressed her laws.— Mr. Jorgensen, though born of respectable parents at Copenhagen, at an early age entered into the British service as an apprentice on board a collier; after which, he employed himself in such other vessels of various descriptions as he thought most likely to promote the object he had in view, that of attaining the highest perfection in seamanship. He then entered our navy, in which, after much hard service and many long and difficult voyages, he made himself complete master of navigation, as well as of the naval laws of Great Britain; and imbibed, according to his own words, together with his knowledge of nautical affairs, the maxims, the principles, and the prejudices of Englishmen. At the age of twenty-five, having been absent from his native country ten or twelve years, the whole of which he had spent in the British service, he returned to Copenhagen in the year 1806. In that city he was at so little pains to conceal his political sentiments respecting England, that he

vided, I proceed without farther preface to the sketch of a revolution, which so far differs from all others of our times, that, in accomplishing it, only twelve men were em-

created himself a number of enemies by declaring his partiality towards a country, under whose flag he had so lately and so long served, and by reprobating in the most open manner the tyranny and usurpation of the French; a nation, whose opinions and principles he found were approved of by the greater part of his countrymen. Shortly after the late expedition, sent by Great Britain against Copenhagen, the Crown Prince entered into an alliance, offensive and defensive, with France; which was followed by a decree, calling upon persons of all ranks and descriptions, from the age of eighteen to fifty, to serve, in some capacity or other, in support of their country; in consequence of which, Mr. Jorgensen took the command of the Admiral Juul, a privateer of twenty-eight guns, in which, proceeding towards Flamborough Head, he fell in with two British ships of war, the Sappho and the Clio. The former he immediately engaged; but, after an action of forty-one minutes, was obliged to strike his colors, and was landed as a prisoner at Yarmouth; whence he was taken to London, where he signed his parole, and remained, till the circumstances, of which the following narrative is intended to convey an account, induced him twice to leave the kingdom, without permission from the British government, and consequently to break his parole;

ployed, not a life was lost, not a drop of blood was shed, not a gun fired, nor a sabre unsheathed.

The island of Iceland, from its climate and situation, and from the exceeding barrenness of its soil, is necessarily compelled at all times to depend for a considerable part of its supplies of provisions upon foreign countries; so that, even in those seasons which may be accounted the most favorable, it does not produce sufficient for the maintenance of its scanty population; and, as often as an unusually severe winter proves destructive to the cattle of its inhabitants, or an unproductive fishery prevents them from laying up their winter stores of dried cod and salmon, nothing but the most abundant imports can avert an actual famine. Such imports in time of peace the parent state of Denmark has found no difficulty in

though he did so, not only without any intention of serving against Great Britain, but, as was shewed by the event in the first instance, and in the second by the proclamation issued in Iceland, with the full determination of returning to England.

furnishing from her numerous ports in Norway, as well as from Copenhagen, but since the breaking out of the present unfortunate war between this country and Great Britain, the naval superiority of the latter has rendered all communication between the former and her colonies most precarious, and the wretched Icelanders have experienced the greatest difficulty in procuring even the poor supplies necessary for their bare subsistence. Sensible of the miserable and defenceless state of this island, it has therefore been the generous wish of the British government that it should be suffered to remain in a state of virtual neutrality, and they have of late gone much farther, and even granted licences to protect vessels belonging to the Danes employed in the conveyance of provisions and other articles of necessity, and to permit English ships to carry similar cargoes thither. "An humane interest," to use the words of Count Tramp, "has been shewn by the English in the fate of the inhabitants, for which they will ever with gratitude remember the exalted philanthropist, Sir Joseph Banks, who on this occasion undertook to advocate their cause."

As yet, however, no commercial communication had taken place between the English and Icelanders, and it unfortunately happened that the first visit they were destined to receive from our countrymen was of a nature but ill calculated to impress them with favorable sentiments towards us; for, benevolent as were the intentions of our government, no public notification had been made of them, and they were consequently of no avail in preventing the depredations of our privateers; one of which, in 1808, under the command of Captain Gilpin, came to the island, and landed an armed force, which took away from the public chest upwards of thirty thousand rix-dollars that were appropriated to the maintenance of the schools and the poor.

Far different from this was the object of Mr. Phelps, an eminent and honorable merchant in London, who, having accidentally learned from Mr. Jorgensen that a large quantity of Icelandic produce, and particularly of tallow, was lying ready for exportation in the ports of that island, conceived the project of opening a direct communication, likely to prove equally beneficial to both parties; and, without delay, freighted a vessel called the Clarence, at Liverpool, for the purpose, in doing which, to avoid all possible cause for umbrage, he, according to Mr. Jorgensen, applied to government for permission to export no other articles but such as were absolutely necessary for the subsistence of the inhabitants, as barley-meal, potatoes, and salt, with a very small proportion of rum, tobacco, sugar, and coffee, not exceeding ten tons; taking especial care not to send out any British manufactured goods, and thereby give room for a charge that he merely wished to make the island a depository for prohibited articles, which might thence be afterwards smuggled into the continent. This ship was furnished with a letter of marque, but still, in order to prove the honorable intentions of the merchant, it was expressly stipulated with the owner, that the captain, Mr. Jackson, should not seize or capture any vessel, either in the ports of Iceland or in sight of its coasts; and in case that he should in any way violate the agreement, the owner should be liable to the forfeiture of £8,000. In this ship

Mr. Jorgensen himself, whose knowledge of the Danish language and general acquaintance with affairs of this nature made him eminently serviceable, embarked, together with Mr. Savigniac, an Englishman employed as supercargo; and, setting out in the latter end of December, they arrived at Iceland in the beginning of the following month, January, 1809; having performed the voyage at a time of the year considered so dangerous for such an attempt, that Mr. Phelps was unable to find any underwriters that would consent to insure the whole of the cargo. The idea having occurred to them that the government of the island would find less difficulty in permitting a free and open trade to be established between the inhabitants and the supercargo, could an appearance be made of the property belonging to neutrals, it was judged expedient to hoist American colors, and to exhibit a set of papers of the same nation; but such an attempt availed nothing, for permission was still peremptorily refused for any part of the cargo to be landed, although it was acknowledged that the country was in extreme want of various articles that were on board. Such being the case, the British colors were displayed, and the licence produced, but to no purpose *; and Mr. Savigniac, unwilling to proceed to extremities, was upon the point of returning to England, when the natives expressed so strongly their anxiety for the landing of the goods, that, in order to bring the government to a sense of its duty and interest, he thought proper to release Captain Jackson from the clause in the charter-party which prevented him from making prizes in Iceland, and to commence hostilities, by taking possession of a Danish brig, which had just arrived from Norway with provisions. The officers of

* Upon the subject of permitting a commercial intercourse, Count Tramp remarks, that, " the existing laws of the country strongly prohibiting all trade with foreign nations, it was the duty of the officers in whose hands he had, at that time, during his absence to Copenhagen, left the management of public affairs, to refuse this application."—It may be so; but, surely, a nation which had conducted itself with so much lenity and forbearance as ours had done towards this island, might have expected to have received a better return for its kind offices.

the government now seeing their real situation, and fearing lest farther acts of a similar nature should be committed, found themselves under the necessity of concluding a convention, permitting a mercantile connection to be opened between the inhabitants of Reikevig and Mr. Savigniac, a measure that in reality was but of little importance, as the natives were still intimidated by the threats of those in power, and dared not purchase of the English; so that every thing went on, as before, through the hands of the Danish factors, who bought only just enough for their own immediate use. How hard this was, will immediately be seen, when it is known that of all the various articles on board the Clarence two only were on any terms to be procured in Iceland, salt and grain, the latter of which was entirely monopolized by government, and not to be purchased at a lower rate than twenty-two dollars per barrel, a price that virtually amounted to a prohibition, as it rendered it quite out of the reach of many even of the higher classes of the inhabitants. Mr. Savigniac, on the contrary, offered his at considerably less than half this price, but still no purchasers* were to be found, nor could he procure even tallow or any other kind of Icelandic produce in exchange for it; so, entirely foiled in his expectations, he was under the necessity of determining to remain himself in the country, and take care of his valuable

* It is but fair to remark, that the time of year, in some measure, prevented so ready a sale of the cargo as Mr. Savigniac seems to have expected, though Count Tramp attributes the failure to a different cause, and asserts that the ship was loaded with luxuries instead of necessaries.—To use his words, " this little town (Reikevig) was now overstocked with luxuries of all descriptions, that could not but find a slow and tardy sale, at a season of the year when the commerce could only be carried on with the inhabitants of the town, and its immediate vicinity; for it is only in the month of June that a degree of communication, intercourse, and barter between the more distant towns and parts of the country begins to take place in Iceland. Of real necessaries, on the contrary, such as the country wanted, and for which there was at that time in particular a demand, only a very inconsiderable quantity was imported, so that, partly owing to these causes, and partly to extravagant sacrifices and expences, and to rash and imprudent speculations and general mismanagements, it was not long before it was reported that this new establishment turned out a losing concern."—*Count Tramp's Narrative.*

charge, hoping for more favorable times, while Mr. Jorgensen returned to England with the Clarence in ballast, having previously restored the Danish brig to her owners.

The governor, Count Tramp, who had been absent at Copenhagen during these transactions, was apprised of them on his return to Iceland on the 6th of June, 1809, and he observes, upon the subject, in his statement, that, "mortified as he felt at a convention of this kind, concluded with an armament unauthorised to enforce it; yet, nevertheless, acknowledging the sacredness of contracts, he had no idea of curtailing in any respect the rights thereby granted to British subjects, though Mr. Savigniac himself, by acting contrary to the convention, had long since given him sufficient cause to have dissolved it."—In the early part of the same month, Captain Nott, of his majesty's sloop of war the Rover, arrived in the country, and an opportunity was thus offered to Count Tramp, as well to prove the sincerity of his intentions, as to render the most essential service possible to Iceland, by fixing all matters in dispute upon a permanent basis with an officer whom he looked upon as no less qualified to enter into an agreement than able to enforce the observance of it *.

On the 16th of June a convention was accordingly concluded between Captain Nott and Count Tramp, by which it was stipulated that British subjects should have a free trade on the island during the war, but that they should be subject at the same time to Danish laws. The governor proceeds in his

* The feelings of the governor I cannot express better than in his own words, "I must beg leave to remark that, from the existing warlike relations, I did not view with indifference the arrival of an armed force belonging to his British Majesty, with the objects of which in these parts I was unacquainted, and the peaceable proceedings of which no convention secured. My duty, therefore, imposed upon me to take every possible means of precaution; but, having been assured that Captain Nott was far from intending any hostility against the country, I could not but wish, under the existing circumstances, that a compact entered into with a man acting under public authority should establish a firm and understood relation between the inhabitants of Iceland and those British subjects who were settled there already, or who might come hereafter for the purposes of trade."

narrative to assure us, that he immediately gave orders that a number of copies of this convention should be printed, and circulated throughout all parts of the island, and moreover that, as this was an operation that required some time, the country affording but one press, which was situated many miles distant from Reikevig, he, still farther to assist its publicity, and to cause it to be known in the vicinity of the residences of the following officers, issued a notification to both the Amptmen in the country, Mr. Thoransen and Mr. Stephensen *, and also gave orders to the Landfogued, Mr. Frydens-

* I should fear that the tardiness of these officers in executing the count's orders must be accounted one of the principal causes of the events he afterwards complained of; for, among other papers with which I have been furnished by Captain Jones, is one containing a narrative of the revolution, by the Etatsroed Stephensen, in his own hand writing, in which, after mentioning the circumstance of the imprisonment of the count, he speaks of the convention with Captain Nott. From this, it does not appear, although he acknowledges the receipt of the copies of the convention, both by him and his brother, that they took any pains to make the contents known in their neighborhood. His words are, "on the same day

berg, and to Mr. Koefoed, a Sysselman, to make the contents public in their neighborhood. The convention was likewise shewn to Mr. Savigniac for his perusal.

(Sunday, the 24th of June) the said convention with Captain Nott, left the press, to which Count Tramp, about six days ago, delivered it, for the purpose of having it printed; and at the same time communicated a copy of the convention to myself and the bailiff Stephen Stephensen, but the count was already confined, before a printed copy could be received by him." It will be remembered, that the convention was concluded on the 16th. The press is indeed a long day's journey distant from Reikevig by land, but it is possible to go by water to within a few miles of the spot, and half a day is sufficient time for the voyage.—Perhaps the disposition of the Etatsroed towards the English cannot be better explained than by giving the following translation of an extract from that gentleman's pamphlet, containing a history of the *Sol* of the Icelanders *(Fucus palmatus)*, printed at Copenhagen, 1808. I have noticed the work in one or two places in my journal, without any reference to the political matters with which it is interspersed.

"*To the good Inhabitants of Iceland.*"

"DEAR COUNTRYMEN!

"Odin's Goddesses, Bellonæ, afflict our northern countries. They have at last shot their murderous and fiery arrows into our king's residence, in a manner

I must here beg leave to observe that, though I would be far from questioning the good intentions and the sincerity of the count, or the correctness of his statements, still, admitting them to be strictly true, some strange neglect had certainly taken place; for the convention had not been printed at the period of our arrival, though five days had elapsed from its signature, the half of which would have been sufficient for the purpose, even supposing it to have been sent by land; and, what is of most consequence, but is omitted in his narrative, a proclamation had been dispersed over the

equally treacherous and shameful. They have, under the mask of hypocrisy, stolen into his country, to rob him of his fleet, and to plunder his kingdom, which was of all in Europe the most happy, owing to nearly an hundred years' peace. They have captured a number of Danish, Norwegian, and Icelandic merchant-ships. With violence and treachery have they provoked the well-merited hatred of our nation, and roused us to arms, in defence of our king, our country, and our liberty. They have surrounded our coasts with ships of war, to destroy our commerce, and to prevent all importation of the most common necessaries into our countries, thereby to the utmost of their power causing misery and the loss of lives."

country, and was found by us still posted up in the town, forbidding any native to trade with the English, under pain of death. This proclamation had been printed during the absence of the count, but kept in a chest till his arrival, and was certainly issued by his special direction.

While things were in this state in Iceland, Mr. Phelps had been planning a second expedition, prepared with more care than the former, and upon a more extensive scale, with the hope of accomplishing his favorite project, and of repairing the losses he had sustained. He therefore, early in the summer, got ready in London the Margaret and Anne, a fine ship carrying ten guns, provided with a letter of marque, and loaded with a cargo of such articles as had been pointed out by Mr. Savigniac as most likely to be saleable, and he, at the same time, dispatched the Flora, a brig, with grain for the use of the island. So much had he this object at heart, that he determined himself to sail in the former of these vessels, to avoid all mistakes, and see that nothing might interrupt the harmony he hoped to find established; and he took with him Mr. Jorgensen, whose acquaintance with the transactions that had taken place during the winter, added to his knowledge of the Icelandic merchants and of the Danish language, with which they are all acquainted, rendered him of the highest importance to the success of the undertaking.

On the 21st of June, only two or three days after the departure of the Rover, the Margaret and Anne came to an anchor in Reikevig harbor, and Mr. Savigniac immediately proceeded on board, where he informed Mr. Phelps of the obstructions * to the trading with the British, of the arrival

* I feel myself bound to mention that Count Tramp, in his narrative, expressly denies any obstructions of this nature, referring to a document, which I have never seen, signed by five of the established merchants of Reikevig, dated the 1st of September, 1809, "by which," he says, "it is evidently proved that the accusations against the civil officers and citizens of the country are unfounded, and that they were invented and set afloat by the managers of the concern, only with a view of providing a cloak for themselves in the account they were going to render to their principals for the losses occasioned by their own misconduct."

of Count Tramp, and of the convention between Captain Nott and the latter.

Mr. Phelps, on hearing that such a convention had been entered into, remained several days without landing any part of his cargo, in the expectation that it would be delivered to him in an official manner, or would, at least, be posted up in different parts of Reikevig. No such thing, however, happened, but, on the contrary, the proclamation before alluded to was still regarded as continuing in force, nor was it ever, at any subsequent time, publicly repealed; so that in the month of June, a season of the year when by far the greater number of the natives make a journey to Reikevig for the sake of barter, scarcely an individual of this description was seen; all naturally dreading to expose themselves to the severe punishment threatened by such a proclamation, and knowing that, except from the English, nothing of what they wanted was to be procured.

Such then being the situation of Mr. Phelps' affairs, and the convention remaining unpublished as late as the 25th of the month, this gentleman felt that longer delay would be materially prejudicial to his interests, and that he must consequently be under the necessity of having recourse to measures, no more consonant to his inclination than to his feelings. He therefore gave orders to Captain Liston, the master of the Margaret and Anne, by virtue of the power granted him by his letter of marque, to seize the person of the governor, and detain him as his prisoner; directing him, also, immediately before he took such a step, to make a prize of the Orion, a brig belonging to Count Tramp, provided with a licence from the British government, which she had, according to Mr. Jorgensen, forfeited, by first carrying her cargo to Norway, and there disposing of it, and taking in another * for Iceland. Mr. Liston, in pursuance of these directions, landed twelve

* A part of this cargo, according to Count Tramp, consisting of goods to the value of six thousand rix-dollars of Danish currency, was intended to have been distributed gratis among the distressed Icelanders, a circumstance of which I am persuaded Mr. Phelps and Mr. Jorgensen were ignorant, or they would not have allowed so benevolent a design to have been frustrated.

of his crew with arms, and, stationing them at the door of the governor's house, entered, together with Mr. Phelps, the room in which he was sitting with Mr. Koefoed, and made him his prisoner, without any resistance on his part: then locking the door of his office, to which he allowed the count to affix his own seal *, he conducted him under an armed escort on board the Margaret and Anne. The whole of this was done without any attempt at concealment in the most public time of the most public day of the week, a Sunday afternoon, after divine service, so that it affords the strongest evidence in favor of Mr. Jorgensen's assertion, that the transaction itself could not be displeasing to the natives, many of whom were loitering about the plain before the house, with their long poles in their hands spiked with iron, which they use for walking upon the snow, and which they might have now employed as offensive weapons; instead of which they looked on with the most perfect indifference, though they were in such

* This was shortly afterwards broken open, and all the papers subjected to examination.

numbers that one half of them could with ease have overpowered the invalids who were stationed to guard the door; for it is sufficiently known, that in time of war the crews of merchant-ships consist of such men only as are unfit for the service of his Majesty. Mr. Phelps, having taken this step, was aware that, as a British subject, he had it not in his power to establish or even to maintain in the island any form of government without the consent of his own; but he was at the same time fully sensible of the necessity of some regular authority being constituted, till more decisive measures could be taken for the welfare of the country; and it was therefore determined that Mr. Jorgensen, not being a subject of the crown of Great Britain, or responsible to it for his actions, should assume for the present the chief command. Conformably to such a determination, this gentleman immediately commenced the exercise of his power by issuing a proclamation*, which in the first

* The extracts from this as well as the two following proclamations published by Mr. Jorgensen I have thought it best to insert almost verbatim from Count

article declares, that all Danish authority is dissolved in Iceland: in the fourth that arms and ammunition of all kinds are to be given up; in the sixth that the keys of private warehouses and shops, money, accounts and papers, belonging to, or concerning, the interests of the king of Denmark or Danish merchants, shall, likewise, be delivered*; in the second third and fifth articles, it orders

Tramp's statement of them, with many of his comments, that I might be the less liable to be accused of partiality; but I have printed literal translations of the originals in the Appendix B. (See Nos. 1, 2, and 5.) Nos. 3 and 4 are copies of other proclamations of less importance, and not noticed by Count Tramp.

* This was preparatory to the confiscation of all Danish property in the island; upon which subject I must beg leave to make use of Mr. Jorgensen's own words. "This," he observes, "was absolutely necessary, for, if such property had been permitted to have been taken away, the country would have been extremely impoverished, since all goods, property, and merchandize on the island belonged to the Danes, as well all that lay in the store-houses, as even what the poor peasant had in the field; for the lower classes were generally deeply in debt to the Danish factors. It was likewise just and proper to detain all such property, whether public or private; for Iceland had certain funds

all Danes to remain within their own houses, and prohibits every one from holding communication with them: the ninth article threatens those who act contrary to this decree with being brought before a military court and shot within two hours; while by the eighth tenth and eleventh articles, are promised to all native Icelanders undisturbed tranquillity and a felicity hitherto unknown *.

in Copenhagen, for which it might be considered a sort of equivalent. Some years ago, a great eruption took place from Mount Hecla, which destroyed a number of people and ruined many. In Denmark and other countries a large sum of money was collected by subscription for the relief of the suffering inhabitants on the island, and deposited in Copenhagen. The sums procured in such a manner have positively never been paid to the Icelanders, but detained by the Danish government. Also, to indemnify Danish merchants for their losses by property confiscated, the court at Copenhagen has nothing to do, but to order them to be paid from the Icelandic funds; and so could the merchants not be sufferers, and there would still remain a surplus, which would more than indemnify government for what public property was seized."

* Perhaps with a view of obtaining his share in the general felicity held forth by this proclamation, a poor

On the evening of the same day, appeared also a second proclamation, proceeding much farther than the other, and decreeing in its first article, that Iceland should be independent of Denmark; and in the fifth, that a republican constitution should be introduced, similar to that which existed before the country was united to Norway in the thirteenth century, but, till this could be

peasant presented a brief to his Excellency, Governor Jorgensen, who favored me with the following translation:

" *A Petition from Biarne Thorlevsen,*

SHEWETH,

" That in the year 1805, my wife Thorunn Gunnlaugdatter was sentenced to two years labor in the Icelandic work-house, only for the simple thing of stealing a sheep, which besides was nothing at all to me. The separation which took place accordingly, occasioned that I was compelled to take a young girl as my housekeeper, who otherwise much recommended herself by her ability and fidelity. The consequence of these circumstances was that the girl produced two little girls, after each other, whose father I am. We were then separated by order of the magistrates, and in this manner must the education of two innocent, but at the same time right handsome little girls, remain neglected, unless she as mother, in conjunction with me as father, settled by the representatives of the people, allowing by the sixth article, to the existing authorities the option of remaining in their respective situations. In the tenth article it is declared that the country shall be put in a state of defence: the twelfth annuls all debts due to Danish merchants in the

is not hindered from following the irresistible dictates of nature in the care and education of the children. But this cannot be done if we are not allowed to marry, and I humbly beg Mr. Bishop Videlin's declaration; so much the more so, as I am convinced of the justice of my cause.

"I also commit my life and worldly happiness to your Excellency's gracious consideration.

"With the confidence and attachment of a subject,

"Biarne Thorlevsen."

Skridnafell within Barderstrand Syssel,
1*st August*, 1809.

To his Excellency, Mr. Jorgen Jorgensen, Protector of the whole Island of Iceland, and Chief Commander by Sea and Land.

As my readers may wish to learn the fate of Mr. Biarne Thorlevsen and his faithful girl, I will add that upon farther inquiry on the part of the bishop into the affair, he found that the wife was anxious for a separation from her husband, when there remained no obstacle to his wishes of entering a second time into the marriage state.

country or abroad, and prohibits clandestine payment of them, under pain of the individual being compelled again to pay the same amount to the new governor: the thirteenth provides against the prices of provisions being exorbitant: the fourteenth takes off the half of all taxes to be levied upon the inhabitants till the 1st of July, 1810: by the eighteenth, all communication with Danish ships is forbidden: the second third and fourth guarantee personal safety and property, and payments of pensions, &c.; while the nineteenth extends the same protection to Danes who do not intermeddle in the political affairs of the island.

It may be observed, that the number of representatives that were to have been sent were three from the southern ampt, one from the eastern ampt, two from the northern, and two from the western ampt.

The government-house was from this time occupied by Mr. Jorgensen, and all public business was as usual transacted in the office belonging to it. The salaries of the various officers under government were also paid; and they so far appeared satisfied * with the present arrangements that none of

* This satisfaction, to judge from the remarks of Count Tramp and the Etatsroed, existed only in appearance, since the former says, "Thus, a new order of things, presenting to view all the miseries that can spring from boundless despotism, was forced upon an innocent people, loyal and faithful to their king. The Danes that had been in public employments, who were now deprived of their places, and laboring under a suspicion otherwise honorable to themselves, of detesting the introduced changes, and meditating schemes for the fall of the usurper, and who were on that account exposed to the same persecutions and ill treatment of which so many instances had been seen, resolved to depart from a country where, with their best wishes, for want of means and assistance they found no possibility of being useful. Many natives in public functions followed their example in resigning, whose offices were filled with the most unqualified persons, by notorious drunkards and flatterers, who were indebted only to their officiousness as spies and calumniators for the favor and protection of the new ruler." —The latter, after mentioning some of the most severe articles in the proclamation, proceeds to affirm that "they did expand a general horror all over the country; and that only the rascality of a few people did approve them, for the gain of money and for the sake of getting some share in the disturbed government or rather in the

the principal ones resigned their situations, though some few * in private expressed their dissatisfaction at the republican form of government about to be established; it seeming to them absurd that an island, to which nature had denied all internal resources, should be proclaimed in a state of independence which it necessarily wanted the means of supporting. The bishop, however, and many of the clergy, at a yearly meeting † of the synod, signed a document, in which they expressed their satisfaction at

high anarchy here. Very indebted officers, being misled by fair promises of more salaries, did submit or approve the altered form, to their lasting shame for having dispensed with that homage and duty they owed to their native soil. Many good officers resigned their situations, the most did not give in the least declaration. We *(Magnus and Stephen Stephensen)*, the Etatsroed and Amptman of the western part of the island, declared our wish to administer our offices only for the present year and according to the laws of our country."

* These persons, however, it must be confessed, were equally averse to their former government.

† Count Tramp asserts that this meeting was attended only by a few of the clergymen of the neighborhood, "who were surprised into a declaration in favor of

the present situation of affairs, and declared their willingness to support it, exhorting all classes of people to do the same.

Many, likewise, of the natives, came forward, conformably to the tenth article of the last proclamation, with an offer of their services, for the purpose of forming a body

Jorgensen, while all the rest in the island remained unshaken in their allegiance;" and he adds in other parts of his narrative that, "though the proclamation of the 26th of June had struck a dread into the minds of people that could not easily be removed, and though means were taken to keep it up, partly by daily scenes of violence and partly by an armed force from the crew of the Margaret and Anne perpetually patrolling the streets, still the new state was by most people considered a bubble, and the public officers in particular, who ought to have been the first to have paid their homage, did not do it, but some laid down their offices, and others declared they would only hold them for the good of the country by virtue of the same authority under which they had hitherto acted!"—He likewise stigmatizes the motives as well as the conduct of those who attached themselves to Mr. Jorgensen, calling them "a contemptible band of idle persons and men of ruined fortunes, attracted by his being beyond measure lavish of the sums of money amassed by his plunder, and by the pompous promises that he daily retailed on paper or held forth in his harangues."

of soldiers; but, for want of a sufficient supply of arms, as, though a search had been made in the houses at Reikevig the day after Count Tramp's deposition, only twenty or thirty old fowling-pieces, most of them useless, and a few swords and pistols had been found, the number of those engaged was necessarily restricted to eight men, who, dressed in green uniforms, armed with swords and pistols, and mounted on good ponies, scoured the country in various directions, intimidating the Danes, and making themselves highly useful to the new governor in securing the goods and property that were to be confiscated. As a farther act of authority, and to shew the clemency intended to be pursued, four prisoners confined in the Tught-huus, or house of correction, one of the most considerable buildings belonging to the town, were released, and the place itself converted into barracks for the soldiers. Some of the troop were soon employed in seizing the persons of two of the civil officers, the Landfogued, Mr. Frydensberg, and Assessor Einersen*,

* See journal, vol. i. page 89, for a farther account of this transaction.

who were kept in confinement, the former for one night, the latter for eight or ten days, both upon a charge of being at the head of a conspiracy to raise a number of men, who were, after securing the English in the town, to have attacked the Margaret and Anne and made prisoners of her crew. The shops and warehouses in Reikevig belonging to Danes not resident in Iceland were from the first day put under guard, and the goods confiscated, and persons were sent to the distant towns to execute the same errand.

Mr. Jorgensen, having now fixed himself in the possession of supreme power, with the title of His Excellency, the Protector of Iceland, Commander in Chief by Sea and Land, posted up, on the 11th of July, another proclamation *, in which it was declared in the first article, " We Jorgen Jorgensen have taken upon ourselves the government of the country until a regular constitution can be established, with power to make war and conclude peace with foreign potentates;" in the second it is stated that

* See Appendix B., No. 5.

the soldiery (consisting as just mentioned of eight natives) had chosen him to be their leader, and to conduct the whole military department: by the third article a new flag is appointed for Iceland, the honor of which Mr. Jorgensen promises to defend with his life and blood: the fourth abolishes the ancient seal of the country and determines that his own private one is to be used until the representatives of the people shall have fixed upon a new one: in the fifth the time granted to the civil officers for declaring their obedience or resignation is prolonged to ten days for the nearest, and four weeks for the most distant parts of the country, after the expiration of which period all who have not given in their declarations are to be suspended from their employments: the sixth article announces that all officers who shall resign are to repair to Westmannoe (Westman's Isles), until an opportunity is found to convey them to Copenhagen: the seventh promises to that part of the clergy who are willing to declare themselves in his favor, that their circumstances shall be bettered: the eighth repeats the intention of placing the island in a state of defence: the ninth announces the design of sending an ambassador to his British Majesty to conclude peace: the tenth contains something relative to the duties and rights of British subjects living in Iceland: the eleventh states that none but Icelanders are qualified to fill public employments: the twelfth shews that Mr. Jorgensen intends continuing in his office until a regular constitution is established: the thirteenth again declares the confiscation of Danish property, which, by the fourteenth, the Amptmend are enjoined to execute: by the fifteenth we learn that some civil officers, in order to secure themselves against the displeasure of the king, their master, had expressed a wish that they might be *compelled* to exercise their public functions: the sixteenth article has for its object the upholding of the new governor by forbidding all irreverence towards his person: in the seventeenth and last it is observed that the laws and regulations shall remain as before until the new constitution is established, except * that it is permitted

* This exception does not at all meet the approbation of Count Tramp, who observes, "that it is very favorable for malefactors and suspicious persons." But the

for every Icelander to proceed from place to place, and to trade wherever and in whatever manner he pleases, without having passports from Amptmend or other authorities; and it is decreed that all sentences and acts of condemnation must be signed by Mr. Jorgensen, before they can be put in execution.

The Icelandic colors* ordained by this proclamation, containing the representation

Etatsroed goes farther, and says that, "the permission granted to ramble without a passport along the country is a circumstance unheard of in other places, and affords very good opportunity to robbers, murderers, troops of thieves, and criminals of all sorts to commit mischiefs and crimes unpunished!" Mr. Jorgensen, however, considers it a just and necessary clause, for, according to the old laws, no person could remove from one district to another without a written permission from an officer; in consequence of which it frequently happened that this officer would not grant a passport, without the peasant promised to buy the necessary supplies for his family from some particular factor, by which he perhaps might be compelled to pay double what would be asked by others.

* The true and old ensign of Iceland is a slit cod or stock-fish, environed by an oval garland.

of three split stockfish upon a dark blue ground, were shortly afterwards for the first time displayed upon the top of one of the warehouses of the town, under a salute of eleven guns from the Margaret and Anne, and were afterwards hoisted upon Sundays, and occasionally on other days. Mr. Jorgensen now, as much perhaps for the sake of finding what merchandise could be procured, as for the purpose of seeing that his various proclamations were respected, accompanied by five of his soldiers, made a journey across the country to its most northern parts, in the course of which he was every where received with the kindest welcome, as well whilst his guard was with him, as on his return when only escorted by a single Icelander. In all places that he visited, the natives crowded about him to relate the impositions they were subjected to by the Danes, and to assure him of their satisfaction in the prospect of being freed from their tyranny.

During the time he was occupied in this expedition, Mr. Phelps was employed in executing a part of his Excellency's orders, by putting the town and harbor of Reikevig in a state of defence, an office he readily undertook for the security of the very considerable property he now had there, as well as of that which he still expected from England. For this purpose a battery, denominated Fort Phelps, was formed near the town, at which the natives, in great numbers, and the crew of the Margaret and Anne, worked with so much alacrity that it was in a short time completed, and mounted with six guns, that had been dug up from the sand on the shore, where they had long been lying; having been sent over from Denmark one hundred and forty years ago.

The order for the confiscation of all Danish property in the island, which was begun to be put in execution immediately after the publication of the second proclamation, was still more vigorously prosecuted on Mr. Jorgensen's return from the north. The property contained in the shops and warehouses in Reikevig, which had from the first day been secured by a guard, was now put under sequestration, and persons were sent for the more effectually enforcing of the decree to the distant factories, such as Havnfiord and

Köblevig. Among other things, possession was taken of two thousand six hundred rix-dollars *, belonging to the public chest, under the care of Mr. Adzer Knudson, and a seizure was made from a Mr. Strube, of a stock of tallow, train-oil, fish, and woollen goods, belonging to a trading company at Flensburg, and another of a considerable quantity

* Count Tramp observes that, according to a specification drawn up by Mr. Phelps, the public money forcibly seized in Iceland by Mr. Jorgensen amounted in the whole to nineteen thousand two hundred and twenty rix-dollars, eighty-six skillings, Danish currency. Mr. Jorgensen, however, who appears to have kept an extremely accurate account of money received either by confiscation or from the public officers, as well as of sums issued in the payment of salaries and for other public purposes, states the former at sixteen thousand nine hundred and fifty-five rix-dollars, two marks, and eight skillings; and the latter at sixteen thousand nine hundred and sixty-one rix-dollars, five marks, and four skillings. Other sums were advanced by Mr. Phelps to meet the demands of various persons, but these did not come under the head of public expences. It is to be remarked, that Mr. Sysselman Koefoed had collected king's taxes to the amount of twelve hundred and ninety-five dollars, which were consequently considered as property to be confiscated; but as this gentleman had laid out the money in the purchase of land, Mr. Jorgensen did not claim any of it.

of goods from a mercantile concern established in Nordburg. I have already mentioned the circumstance of the ship Orion * being made a prize: possession was now likewise taken of the cargo that remained still on board, and the part of it that had been unshipped was also confiscated. It happened shortly after that another Danish vessel, commanded by Captain Holme, which is said by Count Tramp to have had a licence † from Great Britain, arrived in Iceland with a supply of

* This was the only vessel that was seized.

† As a difficulty may be supposed to exist upon the question of licences, and it may be considered by many of my readers that the taking violent possession of a ship furnished with one, must in every case be an act of piracy, I beg leave to subjoin an explanation on this head, with which I have been very lately favored by Mr. Jorgensen. When the British government grants a licence, it is expressly stipulated that the ship shall proceed directly from such a port to such a port, specifying their names. But should it happen, which is very frequently the case with vessels trading to Iceland, that after having procured a licence, in going from an English port they observe the sea clear and free from cruizers, they will run into Norway, sell their cargo there and go back to Copenhagen for another; but if they then, on their way to Iceland, meet an English ship of war, they

necessary articles for the country, the whole of which, together with ten thousand rix-dollars for the payment of the salaries of the public officers, &c., was considered lawful

will produce their licence, though in reality it is no security for that cargo. But should it happen that the people on board the man of war observe such a licenced ship, with a favorable wind, to be steering a course different from her direct one, and thereby deviating from the route pointed out in her licence, that vessel is a lawful prize. At other times, indeed, licences are only granted for a certain limited time, and, if exhibited after the expiration of the period expressed in the licence, such a vessel is also a good prize. One or other was the case with all the vessels in the Iceland ports in the summer of 1809, but none of them would have been condemned in England if they had been seized by the letter of marque, because they were then lying at a port to which their licences permitted them to proceed. That they had forfeited the protection granted them by their licence could not be proved by the ship's papers, though it could from letters to different people on the island: these, however, are not admitted in a court of admiralty. The case of the Orion differs from the former ones, in as much as the person to whom the licence was granted (Adzer Knutzen) was not with the vessel; but since the papers, which proved the forfeiture of the licence, were not on board the vessel at the time of her seizure, she was not considered a legal prize, and was restored to the owner.

plunder, and the Landfogued, Mr. Frydensberg, was compelled to deliver up the public money chest of the country, containing two thousand seven hundred rix-dollars.

In addition to the above, the four following circumstances are stated, as the most aggravating acts of violence and oppression that took place, by Count Tramp, who professes to regard the whole as a regular system of plunder, and considers this as the leading object in every thing that was done by Mr. Phelps or Mr. Jorgensen:—first, that Mr. Savigniac proceeded armed to a settlement at Oreback, belonging to a merchant of the name of Lambertsen*, taking with him a number of horses loaded with goods,

* Of this affair, which is by Count Tramp regarded as a case of peculiar hardship, I have just received from Mr. Jorgensen the following explanation: Mr. Lambertsen is owner of a vessel accustomed to trade between Iceland and Norway, for which purpose, early in the year 1808, he procured from the British government a licence, empowering him to convey to Iceland a cargo of provisions. Of this circumstance he had apprised his factor Sivertsen, who, after waiting till August, 1809, in expectation of the arrival of his principal, concluded that he must either be lost, or that he had taken ad-

consisting chiefly of tobacco and coffee with other articles of luxury, which Mr. Lambertsen's factor was forced to receive, though, far from having ordered any thing of the kind, he

vantage of his licence to carry on an illicit trade, as is often done between the different parts of Denmark, and that he would at all events not appear till the following year. The inhabitants of Oreback being therefore greatly distressed, as two years had now elapsed since any ship had come to them with provisions, and Mr. Lambertsen's own stock of goods, which had been lying all that time in his storehouses, beginning to be injured by keeping, Mr. Sivertsen wrote to Mr. Jorgensen different petitions, begging him to use his interest with Mr. Phelps to supply Oreback over land with things of absolute necessity; to which, after some delay, that gentleman assented, and an agreement was drawn up and signed by both parties, stipulating that a return should be made for goods so sent from Mr. Lambertsen's storehouses. Mr. Sivertsen, in consequence of this, gave an order to Mr. Petreus, Mr. Phelps' agent, for different articles, such as tobacco, coffee, sugar, cloth, &c.; the whole of which was accordingly sent on thirty-five horses, which returned laden with Icelandic produce; the expence of conveyance both to and from Oreback being defrayed by Mr. Phelps. Mr. Lambertsen at this time unexpectedly arrived, and was naturally hurt, as well at finding that the goods with which he had calculated on loading his own ship were in the possession of Mr. Phelps, as that his warehouses were filled with the property of the latter, which would obstruct the sale of what he had

had applied to Mr. Jorgensen to be excused from taking them in; and that, in exchange for these, Mr. Savigniac compelled him to give up a quantity of merchandize, of which a return had previously been made to government:—secondly, that an accusation of oppresion and extortion having been made against Mr. Poulsen, a factor belonging to Mr. Petreus' establishment at Westmannoe, he was, without any inquiry into the grounds of the accusation, or without being allowed to speak in his own defence, summoned to appear at Reikevig, where he was detained for several days, and all the goods upon the island whether belonging to the factory or to any other person were confiscated:—thirdly, a vessel belonging to Mr. Clausen, which had a British licence, was seized and had her cargo confiscated; it was, however, afterwards restored:—fourthly, Mr. Lambertsen,

himself imported. He therefore refused to ratify the agreement, and complained to Captain Jones of what he called the forcible taking away of his goods, requiring to be paid for them, though it was sufficiently notorious that he had already been over paid, in as much as coffee, &c., are more than equivalent to the same weight of Icelandic produce.

the merchant of Oreback just mentioned, on his return to Iceland from Denmark, confident in the security afforded him by his British Majesty's licence, was in like manner ordered to appear at Reikevig, where he was for some time detained, and the cargo he had brought to the country was confiscated. *

* The second and fourth of these charges are of so notorious a kind, that little as I interested myself in political affairs in the island, it struck me while perusing them, that such facts could not well have transpired without their coming under my knowledge; but as I could not call to mind any such circumstances I thought it best to inquire of Mr. Jorgensen how far they were correct. From his answer, which confirms my own ideas, I have a further proof of the inaccuracy of the information which Count Tramp obtained from the Danes and other interested persons in the island, who in these instances have plainly imposed upon him, "and who," to use Mr. Jorgensen's words, "very probably square their accounts at his expence."—The situation of Westmannoe is such, that it never would have answered the purpose to have confiscated property there; besides which the goods belonged to Mr. Petreus, who had none of his property touched either there or at Reikevig. Mr. Lambertsen's cargo was exempt from confiscation; or, had it not been so, there would not have been time to have seized it, since it came to the island but just before Captain Jones' arrival, after whose interference nothing was confiscated. It is true,

An event as unforeseen as it was unfavorable to the present state of political and commercial affairs happened in the arrival at Havnfiord of the Talbot sloop of war, commanded by the Honorable Alexander Jones, to whom the factors of the Danish merchants resident in that place lost no time

Mr. Lambertsen had an order to deliver ten thousand dollars, public money, brought in his ship, but it was never done.—The following information is all that I have been able to procure upon the subject of the revolution in addition to what is related in the first edition of this work. It was communicated to me by my friend, Mr. Clausen, whom I have lately had the pleasure of seeing in England, and whom I particularly requested to point out to me any error or mis-statement which he might find on perusing the narrative.—"I know that Mr. Poulsen was detained some days in Reikevig, charged with oppressing the inhabitants in Westmannoe, which, however, never was proved, and that he was not permitted to speak in his defence, or to get any satisfaction for the improper accusation. But if any goods were confiscated I am unacquainted with the circumstance. During my stay in Reikevig I saw a letter from Mr. Jorgensen, in the possession of Mr. Lambertsen, ordering him not to leave Reikevig without his (Mr. Jorgensen's) permission; which letter afterwards was destroyed by Mr. Jorgensen himself, who obtained it from Mr. Lambertsen under pretence of wishing to peruse it."

in submitting such a partial and exaggerated statement of all that had taken place, as might be expected from men whose passions and whose interests were so materially involved. Captain Jones, therefore, for the purpose of becoming better informed upon this subject, sailed round without delay to Reikevig Bay, where, among the first objects he saw, was the dark blue flag, with three white stockfish on the upper quarter, waving upon one of the warehouses in the town. Immediately upon his arrival, Count Tramp, a prisoner * in the Margaret and Anne, in which he had been confined ever since his

* There appears to me to be no just reason for the severe treatment which Count Tramp states that he received during his imprisonment in the Margaret and Anne. A love of truth and a desire to make the present narrative an impartial one, urges me to the insertion of the count's own relation of these circumstances. Perhaps an apology for indignities offered at the period of the seizure of his person may be found in the hurried manner in which it was done, and the inflamed state of the minds of the persons concerned in it, in consequence of the suspected ill conduct of the governor, but no such excuse can be made in the more tranquil time of the imprisonment, for a filthy cabin and an uninterrupted confinement of nine weeks. With regard to the count's general fare, I always thought that he

capture, solicited an interview with him, when he stated how ill he had been himself personally used, and how contrary to all the laws of nations; adding, that Mr. Jorgensen

was allowed a supply of every necessary from the Landfogued, Mr. Frydensberg, or from his factor, Mr. Simmonsen; and, indeed, I feel almost confident of it.—Yet he says, "Bent down under the weight of so much grief and affliction united, it now became my lot to be kept confined in a narrow and dirty cabin, and sometimes, when Captain Liston took it into his head, even shut up in a small room, or rather closet*, where I was deprived of the light of the day. Constantly I was obliged to put up with the society of drunken and noisy mates, and with them for my companions, I was reduced to subsist on fare which even the men complained of as being more than commonly indifferent; in short, I was deprived for the space of nine weeks of every convenience and comfort of life to which I had been used, and subjected to all the sufferings which the oppressor had it in his power to inflict. His contempt of decorum and humanity even went so far as to refuse a request that was made on my behalf by one of my friends, Bishop Videlin, that I might be allowed to take exercise on a small uninhabited island near which the ship was lying. I would even have submitted to be

* This circumstance happened only once or twice, when the great number of Danes, and the refractory conduct of some of them, called for the assistance of many of the crew from the Margaret and Anne: at such times it was thought the appearance of the count upon deck might encourage the insurrection.

was not only a traitor to his own country (Denmark), but equally so to Great Britain, which he had first served and then fought against; and was now acting in rebellion to both, by hoisting the above-mentioned flag and by declaring the island free, neutral and independent, and at peace with all nations. Captain Jones, in consequence of this information, felt it incumbent upon him to require from Mr. Phelps an explanation of his conduct, and received in answer a brief account of the various transactions which had taken place since his landing in the island, with the motives which had urged

under an armed escort of the ship's crew, if it had been thought necessary, whom I offered to pay for the trouble; yet this request Mr. Phelps refused through Jorgensen, of whose letter to that purpose I have the honor to add a translation. It is remarkable in particular for the assurance it contains that Mr. Phelps could not justify his conduct to his own government, were he to adopt any other measures than those which had been taken."—Even supposing it to be true, as here stated by the count, that he was reduced to live upon the fare of the common sailors, I will not deny that it might appear hard to him who was used to a different mode of living, but I am fully persuaded that such was far from being the case with the sailors, who never had any cause for complaint, nor expressed any.

him to the measures he had adopted *. Having thus far obtained from all parties the most correct information upon the affairs of the island, and having understood from Mr. Jorgensen himself how he was situated with regard to England, Captain Jones considered himself called upon by his official situation to interfere in a business in which the honor of his country appeared to him to be implicated, and he accordingly issued orders that the new Icelandic flag should be taken down; that Mr. Phelps should no longer leave the command of the island in the hands of Mr. Jorgensen, but should, till the will of the British government could be known, either restore the former authorities, or commit the supreme command to some of the most respectable among the inhabitants; that the battery should be destroyed, and the guns taken off the island; that the natives

* A copy of Captain Jones' letter to Mr. Phelps, and the reply of the latter, which contains a more detailed account of what transpired at this time than I have thought necessary to insert in the narrative, will be found in the Appendix B., Nos. 6 and 7. No. 8 of the same Appendix is the copy of a letter from Captain Jones to Admiral Sir Edmund Neagle, explaining in few words his motives for having interfered in the manner he did.

should be no longer trained to the use of arms; that an account of the proceedings should be prepared and dispatched to the British government; and that Mr. Jorgensen and Count Tramp should be forthwith sent to England.

These conditions were accordingly complied with, and an agreement* concluded between Captain Jones and Mr. Phelps on one part, and the Etatsroed Stephensen and the Amptman his brother on the other, in which it was stipulated, that the latter gentlemen, being the next in rank to Count Tramp, should take upon them the government of the island, and be responsible for the persons and property of British subjects. Mr. Phelps, therefore, together with Count Tramp and Lieutenant Stewart of the Talbot (the latter charged with dispatches from Captain Jones) embarked in the Margaret and Anne, and Mr. Jorgensen in the Orion, for England. On the third day of the voyage, however, the Danish prisoners, as is detailed in the journal, set fire to the Margaret and

* See Appendix B., No. 9.

Anne; in consequence of which she was entirely consumed; but the passengers and crew, having been providentially saved by the Orion, returned on the 29th of August to Reikevig, where no other alteration in affairs took place, except that Mr. Phelps and Mr. Jorgensen* with Lieutenant Stewart

* Having thus brought to a conclusion that part of the narrative in which Mr. Jorgensen has been concerned, it may be interesting to some of my readers to know what has since happened to him, and what punishment he has suffered for having unguardedly broken his parole. On arriving in town he took up his abode in his accustomed lodgings at the Spread-Eagle Inn, Gracechurch-street, where, so far from wishing to remain in concealment, he received letters addressed to him without disguise, and even wrote to the Admiralty, and presented himself before the lords commissioners of that court. No notice, however, was taken of what he had done by any of the public offices, until, from private resentment, information was given to the Transport Board that he had broken his parole, and it was farther, though falsely, added, that he had also secreted himself. He was consequently arrested, and confined in Tothill-fields Bridewell, whence he was removed to the usual depôt of prisoners under a similar predicament, Chatham hulks. On board the Bahama, with frequently five and even seven hundred prisoners of the worst description in the same vessel, he was kept in close custody for a twelve-

embarked in the Orion for England, and Count Tramp with his Secretary, and a Dane, a Sysselman of Iceland, (who was considered a necessary witness to the count) were accommodated by Captain Jones in the Talbot.

month. During this interval his bitterest enemies, the Danes, had frequent opportunities of bringing forward charges against him, to which he had no opportunity of replying, but which tended materially to injure him.—He was, however, released from that rigorous confinement, and placed for a while in a comparative state of liberty, upon his parole at Reading; since which time he has been allowed to be fully at large; and here I will beg leave to close my short account of the transactions of this man, by a passage extracted from his manuscript narrative of the revolution of Iceland, which he employed himself in writing during the severity of his confinement.—" If there are any charges against me, let those people making them come forward in an open, fair and candid manner.—Let me see my accusers face to face, and how easily shall I confront them!—but this they dread, for truth must prevail. Where, in the name of God, is there any man in Iceland who can make a just complaint? Is any man injured in property or liberty? Is there any innocent blood crying vengeance against me? If I have shed that of a fellow creature, either in a just or unjust manner, let my head pay for it! If I have gained only one shilling at the public expence, let my right hand suffer for it! If I have enriched myself to the detriment of any one

On arriving in London Count Tramp gladly embraced the opportunity which presented itself of submitting to the under-secretary of state a full detail of all the events that had taken place in Iceland, connected with the late revolution, and a petition for the redress of such injuries as were therein stated to have been received by Danish merchants, or by the Danish government; for it must be remembered that not a single Icelander was injured either in person or property.

What reception these representations met with from our government I have never been able to learn. Certainly no public notice was ever taken of them. To prevent, however, future attacks upon the island from the owners of letters of marque who may be actuated by less honorable motives than

individual, let my left hand be cut off! If I have caused any one single person or more to be confined for being opposite in principles to me, let me feel the horrors of perpetual imprisonment myself! But, if I have done none of these, let me enjoy that liberty which I look upon as the only true good on earth. The British government has a power to crush; it has also a power to be merciful."

those which urged Mr. Phelps to send his vessels thither, Sir Joseph Banks again stepped forward in behalf of his favorite Icelanders, and through his kind and benevolent exertions an order in council was issued, strictly forbidding all acts of hostilities against the poor and defenceless colonies of the Danish dominions, and permitting them to trade with the parent country unmolested by British cruisers. Such conduct on our part could not but give ample satisfaction to Count Tramp, whose own words upon this subject are, "the peculiar favor which Iceland and its concerns have met with here, and the manner in which His British Majesty's ministers have interested themselves in its welfare, and above all the security obtained for the future, has entirely obliterated all bitterness from my heart." In another letter to me he says, when speaking of the proclamation declaring the island to be neutral and the inhabitants placed upon a footing with other friendly strangers, "I apprehend that the people of Iceland with the greatest anxiousness expect the news from England, which, being now so consoling and in every respect so comforting, I should feel myself wanting in duty if I did not forward it as speedily as possible." It may not be improper to insert in this place a copy of the above-mentioned proclamation:

AT THE

Court at the Queen's Palace, February 7, 1810,

PRESENT

THE KING'S MOST EXCELLENT MAJESTY, IN COUNCIL.

"Whereas it has been humbly represented to his Majesty, that the islands of Ferroe and Iceland, and also certain settlements on the coast of Greenland, parts of the dominions of Denmark, have, since the commencement of the war between Great Britain and Denmark, been deprived of all intercourse with Denmark, and the inhabitants of those islands and settlements are, in consequence of the want of their accustomed supplies, reduced to extreme misery, being without many of the necessaries and of most of the conveniences of life.

"His Majesty, being moved by compassion for the sufferings of these defenceless people, has, by and with the advice of his privy council, thought fit to declare his royal will and pleasure, and it is hereby declared and ordered, that the said islands of Ferroe and Iceland, and the settlements on the coast of Greenland, and the inhabitants thereof, and the property therein, shall be exempted from the attack and hostility of His Majesty's forces and subjects, and that the ships belonging to the inhabitants of such islands and settlements, and all goods being of the growth, produce, or manufacture, of the said islands or settlements, on board the ships belonging to such inhabitants, engaged in a direct trade between such islands and settlements respectively and the ports of London or Leith, shall not be liable to seizure and confiscation as prize.

"His Majesty is further pleased to order, with the advice aforesaid, that the people of all the said islands and settlements be considered, when resident in His Majesty's dominions, as stranger friends, under the safeguard of His Majesty's royal peace, and entitled to the protection of the laws of the realm, and in no case treated as alien enemies.

"His Majesty is further pleased to order, with the advice aforesaid, that the ships of the united kingdom, navigated according to law, be permitted to repair to the said islands and settlements, and to trade with the inhabitants thereof.

"And His Majesty is further pleased to order, with the advice aforesaid, that all His Majesty's cruisers and all other his subjects be inhibited from committing any acts of depredation or violence against the persons, ships, and goods of any of the inhabitants of the said islands and settlements, and against any property in the said islands and settlements respectively.

"And the right honorable the lords commissioners of His Majesty's treasury, His Majesty's principal secretaries of state, the lords commissioners of the admiralty, and the judge of the high court of admiralty, and the judges of the courts of vice-admiralty, are to take the necessary measures herein, as to them shall respectively appertain."

(Signed)

"W. FAWKENER."

Hence then it appears that a mercantile speculation the most unfortunate, and a revolution the most singular in its nature, have been the means of placing the island in a greater state of security than formerly; and a way has thus been opened for bettering the condition of its inhabitants, provided the Danish government has compassion enough upon the most injured of its subjects to permit the humane intentions of his British Majesty's ministers to be carried into effect. Should this not be the case (and such seems more than probable, from the late decrees of that country, strictly prohibiting, on pain of death, all intercourse with the British), then will the state of the natives be more wretched than ever; unless, which I sincerely flatter myself will be the case, England should no longer hesitate about the adoption of a step to which every native Icelander looks forward as the greatest blessing that can befal his country, and which to England herself would, I am persuaded, be productive of various signal advantages, the taking possession of Iceland and holding it among her dependencies. Iceland, thus freed from the yoke of an inefficient but presumptuous tyrant, might then, guarded by the protection of our fleets, and fostered by the liberal policy of our commercial laws, look forward to a security that Denmark could never afford, and to a prosperity that the selfishness of the Danes has always prevented: while England would find herself repaid for her generous conduct by the extension of her fisheries, the surest source of her prosperity, and by the safety which the numerous harbors of the island afford for her merchantmen against the storms and perils of the arctic ocean.

END OF APPENDIX. A.

APPENDIX. B.

PROCLAMATIONS, LETTERS,

AND

OTHER DOCUMENTS,

RELATIVE TO THE

ICELANDIC REVOLUTION.

Appendix. B.

PROCLAMATIONS.

N° 1.

PROCLAMATION.

Reikevig, June 26, 1809.

1. All Danish authority ceases in Iceland.

2. All Danes, or factors, connected with Danish mercantile houses, shall remain within doors, and are not to be seen in the streets, nor to converse with each other, nor to send written or verbal messages from one to the other, without having permission so to do.

3. All officers under Danish government shall not leave their respective houses, and are under the same restrictions as those mentioned in the foregoing paragraph.

4. All sorts of arms, without exception, such as muskets, pistols, cutlasses, daggers, or ammunition, shall instantly be delivered up.

5. In case any of the inhabitants, either women or children, shall bring messages to or from a Dane, without permission, they shall be punished as enemies to the state. Nevertheless, should the child be ignorant of its crime, the person sending it shall be punished instead of the child.

6. All keys to public and private storehouses shall be delivered up. All money or bank notes, belonging to the king or factors connected with Danish commercial houses, shall be laid under lock and key. All books of accounts or papers belonging to the king or factors shall be surrendered.

7. Two hours and a half are allowed in Reikevig, and twelve hours in Havnfiord to

execute these orders. Respecting other places, proper arrangements will take place hereafter.

8. All natives, women or children of whatever description, all Icelanders in office have nothing to fear; for they will be treated in the best manner, provided they do not violate the articles contained in the proclamation.

9. Should these orders be speedily executed, it will save a great deal of unnecessary trouble and the effusion of blood. But, on the contrary, should any person act in opposition to what is here directed, he shall immediately be arrested, brought before a military tribunal, and shot within two hours after the offence is committed.

10. Whenever the above articles are known to be carried into effect, a proclamation will be issued, by which the Icelanders will find that nothing but the true welfare of their country is in view, and that our proceedings are solely calculated to insure a peace and happiness little known to the inhabitants in later years.

11. This proclamation shall immediately be translated into Icelandic, and posted up in the most public places, so that the natives may be convinced that nothing will be done prejudicial to their liberty, nor to their disadvantage in any shape whatever.

In case it can be proved that any person shall have acted against the tenor of this proclamation, the person or persons proving the same shall receive a reward of fifty rix-dollars.

(Signed)

JORGEN JORGENSEN.

N° 2.

PROCLAMATION.

Reikevig, June 26, 1809.

1. Iceland is free and independent of Denmark.

2. All public officers, who are natives of Iceland, who remain faithful to their own

country, and who will make oath to execute their functions, shall receive their full salaries.

3. All public officers, who are natives of Iceland, and who remain pacific, shall be respected.

4. All pensions to widows, infants, or officers retired from office, shall be paid.

5. The officers in different ampts or districts shall take care that an honest and sensible person is chosen, who is well acquainted with the situation of his country, and who is to represent his own district. All laws and acts are to come from such representatives. They are to be maintained at the expence of the state, and to be established on the same footing as those were before the island became dependent on the Kings of Norway.

6. Every officer under government, who wishes to remain in his situation, shall notify the same in a letter to me. A fortnight is allowed in the nearest places and seven weeks in the most distant for the purpose, unless bad roads or other insurmountable obstacles

should make it impracticable to furnish an answer so soon; but in such case the reasons for the delay must be communicated at the bottom of the letter. Officers, not remaining in office, cannot expect any assistance from the present government, and other officers will be appointed in their places.

7. None but natives can be members of the legislative body, or represent the people in their respective districts.

8. Iceland has its own flag.

9. Iceland shall be at peace with all nations, and peace is to be established with Great Britain, which will protect it.

10. Iceland shall be set in a state of defence.

11. All hospitals and schools shall be established on a better footing than what they have been hitherto.

12. All debts due to the former Danish government, or the factors connected with Danish mercantile houses, shall not be paid;

nor any money which there is a possibility of remitting to Denmark. Every person so indebted is exempt from paying the sums due. But should any such debtor attempt to pay any part thereof, at any time, he shall be compelled to pay the whole amount to the present government.

13. All kinds of grain shall by no means be sold at exorbitant prices.

14. All Icelanders are exempted from one half of their taxes till the 1st of July, 1810.

15. The inhabitants can proceed uninterruptedly from place to place and trade wherever and with whomsoever they please, except Danish merchants not resident in Iceland.

16. Till such time as the natives shall send in their representatives, all public officers, and persons who have money due from government, may address me for the same, and they shall be paid.

17. No man shall be judged or punished, after the representatives are assembled, with-

out being found deserving of punishment by twelve of his fellow-citizens.

18. Every public officer, of whatever denomination, shall endeavour to prevent all communication with Danish ships. A specification shall be given in of what grain is necessary for each district, so that it may be transported thither before the setting in of the winter; and care shall be taken that hereafter the island shall be supplied with corn for one year to come.

19. No Icelander must, on account of the late liberty being granted, presume to offend or assault a person for being a Dane, nor for having held a situation under the king, nor for having been in the employment of, or connected with, a Danish mercantile house, provided they do not interfere with the political affairs of the island.

*** The number of representatives to be sent are three from the southern ampt, one from the eastern ampt, two from the northern, and two from the western ampt.

(Signed)

JORGEN JORGENSEN.

N° 3.

PROCLAMATION.

Reikevig, June 29, 1809.

We are informed that certain evil-minded people have propagated false reports in the country; and have represented to the inhabitants that it is dangerous to travel from place to place, and that much blood has been spilled in the streets of Reikevig by the English. The inhabitants need not be under any apprehensions, but may rest assured that no violence will be committed against them, and that they are at full liberty to follow their lawful occupations without molestation; and it is hereby declared that all such rumours are entirely without foundation.—All persons that do or shall hereafter spread such false reports, shall be deemed enemies to the state, and it will be necessary to treat all such people, who do not demean themselves as peaceable citizens, with the utmost severity.

(Signed)

JORGEN JORGENSEN.

N° 4.

PROCLAMATION.

Reikevig, July 1, 1809.

We are informed that some discontent exists on account of the natives interpreting an article in the proclamation of the 26th of June, to a total exemption from all debts whatever.—It is hereby declared, that only such debt is remitted which is due to the king, or to such Danish mercantile houses, whose principals are not residents of Iceland. It is further declared, that all such Danish merchants, whose wives and children are at present in Iceland, and who themselves wish to remain in the island, shall receive all debts due to them, and, in case of refusal, the persons concerned will communicate the same to me, who engage to see justice done. On the other hand, all such natives as have money due from Danish merchants' mercantile houses on the island, shall have the same paid to them, if such debt can be proved by their books.

(Signed)

JORGEN JORGENSEN.

N° 5.

PROCLAMATION.

Reikevig, July 11, 1809.

In our proclamation, dated the 26th of June, 1809, it was requested that the nearest districts should, within a fortnight, and the more distant, within a certain limited time, send in representatives, to consult what was best to be done in the present exigency. We find, however, that the public officers have far from facilitated such a meeting; and we are therefore under the necessity of no longer resisting the wish of the people, who have earnestly solicited us to manage the administration of public affairs, and who have in hundreds offered to serve in the defence of their country.—It is therefore declared,

1. That We, Jorgen Jorgensen, have undertaken the management of public affairs, under the name of PROTECTOR, until a settled constitution can be fixed on, with full power to make war or conclude peace with foreign powers.

2. That the military have nominated us their commander by land and sea, and to regulate the whole military department in the country.

3. That the Icelandic flag shall be blue, with three white stockfish thereon, and the honor of it we promise to defend at the risk of our life and blood.

4. That the great seal of the island shall no longer be respected; but that all public documents of consequence shall be signed by my own hand, and my seal (J. J.) fixed thereunto, until such time as the representatives shall assemble and provide a proper seal.

5. That all public officers, who have, from motives of patriotism, already given in declarations that they were willing to serve their country in its late difficult and dangerous situation, shall receive their salaries.—On the contrary, those that have been situated near Reikevig and not yet declared themselves, are totally suspended from office, pay, and power, unless they within the 20th of

this month give proper reasons for not having sent in either their resignations or their wish of continuing in office. After that date a list shall be made out and publicly distributed, of the names of all those officers who shall either resign or continue in their employments. Any person from the date thereof, who shall obey any order from such persons as have not declared themselves shall be deemed a traitor and treated accordingly. Nevertheless a month is granted to persons residing in places more remote from Reikevig, that they may have sufficient time to send in their declarations.

6. That all officers who wish to resign shall be sent to Copenhagen free of expence, when an opportunity is found so to do. In the mean time we command that all such officers shall hold themselves in readiness to be removed to Westman's Isles, so that they may not by their intriguing disturb the public peace and tranquillity, unless they can give security for their future good behaviour.

7. That we have seen with the greatest satisfaction that the Icelandic clergy, as good

christians, have promoted tranquillity and good order at this dangerous period; therefore we promise to pay all their salaries and pensions to clergymens' widows, and also to improve their present situation as much as possible.

8. That the country shall be set in a proper state of defence, without additional taxes on the nation.

9. That a person shall be invested with full power to conclude a peace with his Majesty, the King of Great Britain.

10. That all British subjects shall have full permission to trade and reside in this country, in case they do not offend against its laws; and all who shall unprovokedly assault a British subject, shall be punished.

11. That none but natives can hold either civil or clerical offices.

12. That we declare and promise to lay down our offices the moment that the representatives shall be assembled. The time

appointed for the convocation of the assembly is the 1st of July, 1810; and we will then resign when a proper and suitable constitution shall be fixed on; and it is declared that the poor and the common people shall have an equal share in the government with the rich and powerful.

13. That all Danish property on the island shall be confiscated for public use; and if any one shall conceal money, or other Danish goods or merchandize, he shall be punished.

14. That the Amptmend, whether they remain in office or not, shall see these our orders duly executed, and shut up and put seals on all Danish storehouses in their ampts, and receive all confiscated monies.

15. That several officers, from fear of the Danish government, wish to be forced to retain their offices, though they fully approve of our late proceedings, and therefore do we declare, as we have nothing in view but the real good of the country, that all such people, as are not animated by sufficient patriot-

ism to serve their own country, are permitted to leave the island and go to Copenhagen.

16. The situation we now are in requires that we should not suffer the least disrespect to our person, neither that any one should transgress the least article of this our proclamation, which has solely in view the welfare of the inhabitants of this island. We therefore solemnly declare, that the first who shall attempt to disturb the prosperity or common tranquillity of the country shall instantly suffer death, without benefit of the civil law.

17. In all other respects the ancient laws and regulations shall remain in full force till such time as the constitution is settled, with the exception that every Icelander is permitted to proceed uninterruptedly from place to place, and to trade wherever and in whatever manner he pleases, without having passports from Amptmend or other authorities; yet all sentences and acts of condemnation must be signed by us before they can be executed.

(Signed)

JORGEN JORGENSEN.

Mr. Jorgensen's seal (J. J.) is affixed to the original of all these five proclamations.

Mr. Jorgensen, in his own narrative, remarks as follows : " Many have found fault with different articles in my proclamations, and alledge, that they were written with a great deal too much severity: but this proceeded from a perfect knowledge of the people I had to deal with. Even if there had been some inconsistency in them, such could not be wondered at, for we are not to look for the same regularity, during the period of a revolution, as when a country is perfectly tranquil. But expressions of severity were absolutely necessary on my part, thereby to keep the unruly in check; for I knew my own temper so well, that, had the success of my undertaking depended upon the shedding the blood of one single of my fellow creatures, I should have been obliged to desist entirely; so that, by appearing what I really was not, I managed the whole island with ease. Although it was said in the proclamations, that all officers who did not remain in their situations should be transported to Westman's Isles, such a thing was never attempted to be put in force, neither was a single individual who resigned, driven out of his habitation, which might have been done in those houses belonging to government; but, on the contrary, I ordered that such persons and their families should be supplied gratis with every necessary from the public stores, till an opportunity offered for them to be sent to Denmark. Moreover, though it was publicly proclaimed, that any one, who should disturb the public tranquillity and not deliver up their arms, should be severely punished, the people so offending were only slightly reprimanded. John Bergman, with a drawn cutlass, ran about Reikevig, threatening destruction to us all, for which he was only confined for two hours. Mr. Finböge, who had concealed two thousand six hundred rix-dollars, belonging to Adzer Knutzen, received no manner of chastisement. Sigurd Thorgrimsen suffered no punishment for propagating reports about the country that the streets of Reikevig were stained with the blood of Danes and Icelanders. Assessor Einersen was arrested and confined for a few days, upon the information of the Etatsroed Stephensen, that he was at the head of a conspiracy raised to attack the English.—So much for my cruelty and severity."

N° 6.

COPY OF A LETTER

FROM CAPTAIN JONES TO MR. PHELPS.

His Majesty's Sloop Talbot, Reikevig Bay, August 19*th,* 1809.

SIR,

The conversation I had with you respecting your transactions on this island was not with any intention of interfering, or depriving you of any power granted by government; but, in consequence of hearing reports respecting your conduct, I deemed it my duty to ascertain the nature of your situation, your business here, by what authority you acted, and how far these reports were correct, in order to give you every protection and assistance as a British subject. You having, however, declined in the first instance (deeming it unnecessary) to give me such information, and then sending me a statement of your conduct, which appeared to me in several respects not sanctioned by your having a letter of marque only, I acquaint you, not only that such a communication was a respect due; but that it is a duty incumbent on all British subjects to give every information, both relating to themselves and others, to the captains of any of his Majesty's ships; and your being apprised by letters from the lords commissioners of the admirality of my being sent here to protect your trade and that of this island, ought to have convinced you of its necessity. I therefore sent a message, appointing an hour for you to wait on me, in order to gain such information as would direct me how to act; or, if necessary, to reply in writing to your statement; which circumstances, together with your refusal to comply with my request and a copy of this my answer, I shall transmit by the first opportunity to the right honorable the lords commissioners of the admiralty. I also conceive it my duty to acquaint you, that from your not having any other authority, that I am aware of, besides being owner of a letter of marque, you appear to me to have far exceeded that authority by taking on you the government of an island not actually considered hostile to Great Britain; the wretched state of whose inhabitants his Majesty has been graciously pleased so far to relieve in winter, as to grant licences to you and even to the enemies of Great Britain to support them; and you have, in my opinion, not only transgressed the laws of Great Britain, but of all nations, by assuming an authority which no subject of any realm whatever can have a right to; namely, that of declaring the island free, neutral, independent, and at peace with all nations, and of appointing a governor, who is not a British subject, but a Dane; who has been an apprentice on board an English collier; served his time as a midshipman in his Majesty's navy; afterwards fought against Great Britain; and was made a prisoner by an English ship of war. I understand he has issued, with your sanction, proclamations (declaring the island no longer under the government or control of Denmark) signed in a regal manner (We, Jorgen Jorgensen); besides which, he has, in sight of his Majesty's ship under my command, hoisted a flag as yet unknown: and is employed at this time in erecting a battery within musket shot, without my permission, and even without having consulted me on the subject; which is not only taking up arms against his own country (Denmark), but a disrespect to my pendant. I feel myself called upon, therefore, to notice his conduct, which no attachment or zeal that gentleman may have for Great Britain can countenance; neither would it, I am sure, meet the approbation of government. I now most earnestly recommend, either that you do not leave the whole power in the island in the hands of that gentleman alone, until his Majesty's pleasure is known (however qualified or respectable his character may be), or, that you immediately restore the former mode of government, giving the supreme command to some of the most respectable of the inhabitants of the island. It was indispensably necessary, and was your duty, long since to have sent an account of your proceedings to government, which I now recommend you not to delay. I also recommend your destroying the battery now erecting, taking the guns off the island, and desisting from training the inhabitants to arms, which can only tend at present to their disadvantage, they being still the subjects of our enemies, therefore not liable to be attacked by those, and may hereafter enable them to turn against Great Britain. Nor can I conceive from the statement you have made of their attachment to you, that such measures are at all necessary, either for the safety of your person or property.—Having thus, according to my duty, acquainted you with my sentiments, and pointed out the line of conduct that I conceive you, as a British subject, ought to adopt, I shall not interfere farther than by requesting to be acquainted with your future intentions, for the information of the right honorable the lords commissioners of the admiralty. You are wrong in supposing that I wish to cast any stain upon your character, either as an Englishman or a man of honor, nor can I believe you would intentionally commit an act which would reflect disgrace upon the British government. I am also far from doubting the word of Mr. Jorgensen, or from throwing any reflections either on his former situation, his character, or conduct; but his not possessing any written document to certify that he has permission from government to be on this island, and his having appeared in Havnfiord Bay, on board his Majesty's ship under my command, in the undress uniform of a post captain, oblige me to insist on his immediate return to Great Britain, unless you can satisfy me you have permission to bring him here.

I am, SIR,

Your most obedient and humble Servant,

ALEXANDER JONES,

Captain of His Majesty's Sloop, Talbot.

To Samuel Phelps, Esq., English Merchant,
Reikevig, Iceland.

N° 7.

COPY OF A LETTER

FROM MR. PHELPS TO CAPTAIN JONES.

Reikevig, August 23rd, 1809.

SIR,

The convention or agreement, which we entered into yesterday with the chief justice and the bailiff of this island, will, perhaps, preclude the necessity of my giving you a circumstantial account of every particular and minute transaction which has taken place here since my arrival, of which I have kept a regular journal, for the purpose of laying the same before his Majesty's ministers, together with original documents. The accounts and papers are voluminous, and it would take a considerable time to copy them: it was, moreover, impossible to do it in the time required by your letter of the 20th instant. The journal papers and documents I hold are necessary for my justification, and it may be of considerable utility to me to retain them; but, as I am totally unacquainted with the laws and articles of war (farther than what I learn from having read the Margaret and Anne's letters of marque), I will readily deliver up all these papers and journal to you, if you will give me an order so to do, and a receipt for the same; as will also Captain Liston his journal and papers.—I must, however, beg leave to correct some errors or mistakes, which appear to exist, according to your letter of the 19th instant.—As to my having declined giving you a statement of my transactions here, this I certainly did not intend; but, as the charges made against me from common report only, as stated in the first conversation I had the honor to hold with you, were of a serious nature, and such as I knew to be unjust and untrue, I wished the whole transactions and complaints to be stated in writing, to prevent misinterpretation. Perhaps this request of mine was not correct or consistent with the respect due to a British officer; if so, I can assure you no such disrespect was intended.—I took the liberty of writing you a letter (in haste) dated the 16th, and another the 17th instant. Of the first I had not time to take an exact copy; but it appears by your letter of the 19th instant, that some words in one particular passage were left out, which were intended by me to have been added. I refer you to the public proclamations, to prove to you that the error in my letter proceeded from the hurry of writing: therefore I must beg leave to correct it.—You state in your letter of the 19th instant, that I have not only in your opinion transgressed the laws of Great Britain, but of all nations, by assuming an authority which no one has a right to assume, namely, that of declaring the island "free, neutral, and independent:" here should have been added, "of Denmark;" for so the proclamation is translated to me.—The only hostility I have committed is against the Danes.—You will find, Sir, by every true information you can obtain on shore, that I have never in any respect interfered in the government or change of government here, farther than by giving my advice and consent to Mr. Jorgensen in matters in which my trade was concerned, or in measures that I was bound to pursue, according to the instructions in the letters of marque, or to instruct Captain Liston so to do; namely, not to compromise in any manner with our enemies. As far as I have gone, I shall not have the business to retract, whatever may be the consequence to me, and sure am I no one proof or document can appear to shew that I have in any way interfered in the government; but I beg leave briefly to state to you how far I have been concerned, and to add at the same time that, not understanding the Danish language, it is possible that I may in some instances have been deceived.—In January last, myself and my partners sent a cargo of provisions and other necessaries under a British licence, to relieve the inhabitants of Iceland. The cargo was landed, but, through the artifice and intrigue of the Danes, instead of returning a cargo of Iceland produce, as the licence specified, the vessel was returned in ballast with stones which our agent was obliged to pay for, although the then constituted Danish authorities had granted us a free trade, and the warehouses were full of Iceland goods. Severe proclamations were also afterwards published to obstruct our trade, all of which I shall take home.—On finding that the same conduct prevailed on my arrival here with another cargo on the 21st of June last, and that I must again return in ballast, unless I pursued strong measures, I ventured to make Count Tramp prisoner, partly on this account, and partly on hearing that he had come here under a fictitious name and character. Being requested by many of the native inhabitants to issue some proclamations, to satisfy the minds of the people, and being also requested and entreated by them to remove the Danes from the island, who had reduced them to the greatest state of misery, I declined interfering, or taking any part in the government, and refused to hoist the English flag, not knowing that I should be correct in so doing, until the will of his Majesty's ministers could be known.—Upon farther applications being sent to me, which expressed the wishes of the people that Mr. Jorgensen would stand forward to protect the island and the natives against the Danes, I certainly acquiesced, and gave him my concurrence and advice to act as the natives thought best, unless some of them would take the management of the island *(ad interim)* till the will of the British government could be known; and I agreed that the island should be declared free and independent of Denmark, but only in suspension, until his Majesty's pleasure could be known. I have in no other respect interfered in the government here, farther than in protecting our property and persons. It has also been impossible for me to transmit an account of these transactions before to England, as there have been no vessels fit for the navigation of these seas; of which you can easily assure yourself. You will find, Sir, that there are two parties in this island; the Danish party and the Icelandic, or English, party. I hope I have listened to and favored that party which will be most approved of by his Majesty's ministers; but, if it should not prove to be so, I have erred unintentionally.

I have the honor to be,

SIR,

Your most obedient and humble Servant,

SAMUEL PHELPS.

To the Honorable Alexander Jones,
Captain of His Majesty's Ship
Talbot, Reikevig Harbor.

N° 8.

COPY OF A LETTER

FROM THE HONORABLE CAPTAIN JONES TO REAR-ADMIRAL SIR EDWARD NEAGLE, COMMANDER IN CHIEF, *&c., &c., &c.*

His Majesty's Sloop, Talbot, Reikevig Bay, Iceland, August 23rd, 1809.

SIR,

I have to acquaint you that, owing to extremely bad weather, I had no communication with this island until the 5th of August, when I anchored for a few hours in Oreback Bay. I was there informed that there were three English vessels at this place and that an English merchant had become governor of the island. On the 8th I anchored in Havnfiord Bay to water, when I learned that Mr. Phelps, an English merchant, and the owner of the Margaret and Anne letter of marque, had made Count Tramp, the Danish governor of the island, prisoner, and that Mr. Jorgensen (a Danish gentleman brought from England by Mr. Phelps) had taken upon him the government of the island; but that it was not satisfactory to the inhabitants, in consequence of his former situations in life. I came here on the 14th from Havnfiord, for the purpose of being better informed upon this subject, as well as for the sake of repairing my rigging. On entering the harbor I found at anchor the Margaret and Anne privateer with two or three other vessels, and saw a blue flag, with three white fish in the upper quarter, flying over the town: this I was told was the new Iceland colors. Count Tramp, who was a prisoner on board the privateer, requested to have an interview with me, during which he stated that he had been extremely ill used, and in a manner that was contrary to the laws of nations; that Mr. Jorgensen was a traitor to his own country (Denmark); that he had first served Great Britain; then fought against it; and was now, by hoisting the above-mentioned flag, and by declaring the island free, neutral, and independent, and at peace with all nations, acting in rebellion to both. I therefore thought it necessary to inquire of Mr. Phelps by what authority he acted. That gentleman having first declined to give any explanation to me, and afterwards sent me the enclosed written statement of his transactions, together with my being informed by Mr. Jorgensen himself that he had been an apprentice on board an English collier, served as a midshipman in the English navy, then commanded the Admiral Juul Danish privateer, which was captured by his Majesty's sloop Sappho, that he was not exchanged, neither had he signed any parole of honor as a prisoner of war, but was set at liberty in London without any written documents: all these circumstances considered, I deemed it my duty to prevent his being left alone on the island at the head of a government of his own formation, and have consequently taken those steps to obviate it, which to the best of my opinion and that of my officers would be right and most approved of by government. Enclosed I send copies of my answer and order to Mr. Phelps, in consequence of the before-mentioned statement, also every other document in my power to obtain, hoping they will give every information and explanation necessary on the subject. Not considering myself authorised to interfere with the imprisonment of Count Tramp, nor thinking it would be proper under the present circumstances to hoist the British colors on the island without first informing you; at the request of Mr. Phelps, I have restored the former government to the two men next in rank on the island to Count Tramp. In order to secure the quick and safe delivery of this intelligence, I have sent in the Margaret and Anne Lieutenant Stewart of his Majesty's ship under my command, who is acquainted with all the circumstances, and charged with the delivery of the new Icelandic flag to you.

I have the honor to be,

SIR, &c.,

ALEXANDER JONES.

Rear-Admiral Sir Edward Neagle,
Commander in Chief, &c.,&c.,&c.,
Leith.

N° 9.

COPY OF AN AGREEMENT

Between His Royal Danish Majesty's Counsellor of State and Chief Justice of Iceland, and His Danish Majesty's Bailiff in the Western County of the said Island on the one part, and the Honorable Alexander Jones, Captain of His British Majesty's Sloop of War, Talbot, and Samuel Phelps of the City of London, Esq., on the other part.

ARTICLES.

1. All proclamations, laws, appointments, &c., made by Mr. Jorgen Jorgensen, since his arrival in this country, are to be abolished and totally null and void, from the moment this agreement is signed.

2. The former government is to be perfectly restored, and the chief command to devolve upon the said chief justice of Iceland, and the said bailiff of the western county of Iceland, native Icelanders, they being the next in power in the island to Count Tramp.

3. All officers under the Danish government are at liberty to return to their offices.

4. The government shall be responsible for the protection of all British subjects and the property that now is and may be on the island, and all transgressions, thefts, and personal assaults, committed against British

subjects or their property, shall be punished with the same rigor, and according to the same laws, as if the property belonged to the natives.

5. No battery is to be erected; and the one now at Reikevig is to be destroyed. No militia is to be raised in the island, nor the country in any way to be fortified or armed.

6. All officers or other persons either armed or unarmed, who, during the late events, have taken part with Mr. Jorgen Jorgensen, shall no longer be in employment, but their persons and property in every respect (whosoever or of whatever nation they may be) shall be respected and protected the same as other persons and natives.—The convention between Count Tramp and Captain Nott, of the 16th of June last, shall be in full force, and be published throughout this country without delay, together with this agreement.

7. All merchants' houses which are shut up in this country shall immediately be opened and the merchants of the said island

be permitted to continue or carry on their trade as formerly.

8. All Danish property and public money is to be restored.

Witness our hands and seals this twenty-second day of August, one thousand eight hundred and nine.

(Signed)

ALEXANDER JONES.
SAMUEL PHELPS.
MAGNUS STEPHENSEN.
STEPHEN STEPHENSEN.

END OF APPENDIX. B.

APPENDIX. C.

ACCOUNT OF HECLA;

WITH

SOME PARTICULARS OF OTHER REMARKABLE

VOLCANIC MOUNTAINS,

IN ICELAND.

Appendix. C.

ACCOUNT OF HECLA,

&c. &c.

There is scarcely a part of this great island but bears the most striking marks of the effects of subterraneous fires, though the more dreadful eruptions of later years have been confined to its southern and eastern quarters. Not only in the loftiest mountains, but even in the plains and vallies, the remains of extinguished craters arrest the attention of the traveller, and the tracts of lava which he frequently meets with are so extensive, that it not uncommonly happens that an Icelandic summer's day, however considerable its length, does not allow sufficient time to traverse one of them in its greatest extent. Among the numerous volcanoes, Hecla, from the frequency of its fires, from its vicinity to the most populous part of the island, and from its situation that renders it visible to ships sailing to Greenland and North America, has been by far the most celebrated among foreign countries; nor does it seem to have been considered of less importance at home, having attracted to such a degree the notice of the native historians, that its several eruptions, subsequently to the colonization of the island (for it is supposed that many had taken place previously), have been inserted in the chronicles of the country. Upon this subject, however, the different annalists are far from being agreed, some, according to Povelsen and Olafsen, who speak only of the principal ones, mentioning no more than eleven, and others only sixteen; while these authors say that, after the most attentive researches, they can speak with confidence to twenty-two, without reckoning several, which, though enumerated by other writers, they regard as uncertain, thinking that the same eruption may have been counted more than once, from its having lasted above a year, or its having begun in winter and ceased the following spring; and also without including the less important discharges, that have not issued from the volcano itself, but from some of the hillocks or beds of lava about it; though these in reality have a right to be included, as having originated from Hecla, whose fire passing through subterraneous channels has found vent in different places. Leaving this question undecided, I confine myself to Arngrim Jonas, who, in his *Brevis Commentarius de Islandia* *, relates the first to have taken

* This account does not exactly agree with that given by Von Troil, who mentions eruptions of Hecla, in 1374, 1390, and 1436.—The dates of the eruptions of Ætna and Vesuvius have also been recorded, and, in the uncertainty of their periods, resemble what I find respecting Hecla.—They are as follows:

Mount Ætna—before the christian æra, four;—in the years 3325, 3538, 3554, 3843.—After Christ twenty-seven—1175, 1285, 1321, 1323, 1329, 1408, 1530, 1536, 1537, 1540, 1545, 1545, 1554, 1556, 1566, 1579, 1614, 1634, 1636, 1643, 1669, 1682, 1689, 1692, 1747, 1755, 1766.

Mount Vesuvius;—after Christ—79, 203, 472, 512, 685, 993, 1036, 1043, 1048, 1136, 1506, 1538, (the

place A. D. 1104; and to have been succeeded by others in the years 1137, 1222, 1300, 1341, 1362, and 1389, after which the mountain is said to have remained quiet till 1538, and then again for the space of eighty-one years, when, in 1619, fresh matter was vomited forth; and also in 1636, 1693, and 1766; the latter eruption lasting, without intermission, from the 15th of April till the 7th of September. Flames, but unattended with lava, appeared in 1771 and 1772, since which period to the latter end of the year, 1810, neither fire nor smoke has been perceived.

Having already, in my journal (vol. i. page 194) stated the circumstances which prevented me from reaching Hecla, it is necessarily out of my power to give an account of the state of the mountain from my own actual observation; but, if I may be allowed to judge from the information I received in the neighborhood, I had less reason than might be imagined to regret my

eruption at Puzzole), 1631, 1660, 1682, 1694, 1701 1704, 1712, 1717, 1730, 1737, 1751, 1754, 1760, 1766, 1767, 1770, 1771.—*Sir William Hamilton's Campi Phlegræi, p.* 51.

disappointment; the covering of snow, that in many seasons entirely envelopes the summit, having lain particularly thick during the summer of 1809, and so completely concealed every thing that might be looked upon as remarkable, that the prosecution of my journey would but have added to my fatigue without a chance of the success I wished for. Sir Joseph Banks, however, and his party, were more fortunate, and an account of their expedition has been published by Von Troil, whose remarks on Hecla are so familiar to the English reader, that the mountain may be considered as well known. At the same time, as it is one of those things that are reckoned most wonderful in Iceland, I am unwilling to pass it in silence, but shall endeavor, by means of extracts from the less generally known publication of Povelsen and Olafsen, aided by some notes made from Sir Joseph Banks' manuscripts, to compensate for what I have not in my power to relate in my own journal.

Our Icelandic travellers, on their excursion to Hecla, stopped at the village of Selsund situated in the vicinity of the mountain, where the proprietor of the farm urged them to accept him for their guide, he being acquainted with the country all around the volcano, though he had never actually reached even its foot. The whole of the inhabitants who reside in the neighborhood consider it as the height of temerity for any one to endeavour to climb the mountain: in order, therefore, to deter these gentlemen from being rash enough to make the attempt, they represented a variety of supernatural obstacles, which, having, from time immemorial, been handed down from father to son, were perhaps as devoutly believed as they were seriously related, telling, among other things that were also urged to me, how Hecla is guarded by a number of strange black birds resembling crows, but armed with beaks of iron, with which they would receive in a very ungracious manner any man that might presume to infringe upon their territory. The country for two leagues around Hecla they found wholly destitute of vegetation, the soil consisting of scoria, pumice, and red and black cinders, which, by the breaking out of the subterraneous fires, were here and there raised into numerous little hills and eminences, increasing in size the nearer they approached the mountain. The principal one, which is called Raud-oldur *, is of an oblong form, with an opening in its summit of an hundred and forty-four feet in depth, and eight hundred and forty feet in circumference: it consists entirely of small red shining stones, that have evidently been in a state of liquefaction. On reaching Hecla, the difficulty of proceeding was increased, especially when it became necessary to travel over the heaps of lava that have flowed from the volcano, and formed round the base of the mountain a sort of rampart from forty to seventy feet in height, consisting of masses

* "We arrived (September 24, 1772) at a green spot under Graufel-hraun where we pitched our tents and proceeded to a crater which has an opening of half a mile in circumference, but its western side is destroyed by the eruption. The hraun lies as if it came from this crater, and the tufa and ashes which formerly made a part of its western side are still seen among it. The lower part and remaining walls are composed of nothing but ashes, cinders, and pieces of lava in various states. Its name is Röd-Oldur.—The scene of desolation all around is almost inconceivable."—*Sir Joseph Banks' Manuscript Journal.*

of melted stone. In this spot, which appears to be the place alluded to by Von Troil, where he speaks of the hill as being surrounded with lofty glazed walls, and filled with high glazed cliffs not to be compared to any thing he ever saw before, our travellers found it necessary to leave their horses; and their guide, under the pretence that he was suddenly attacked with a head-ache, excused himself from attending them farther on their journey. The ridge of lava was climbed with extreme difficulty, for the stones of which it was composed lay detached, and there were so many deep holes between them, that it was necessary to use the greatest caution in walking to prevent accidents. The ground shortly after becoming more solid, their road was consequently materially improved, and they began their ascent on the western side, where the continual cracking of the rock under their feet at first caused them some uneasiness, till, upon more attentive observation, they found that the whole mountain itself was reduced to a mere pumice-stone, lying in horizontal strata of moderate thickness, every where full of fissures; and hence, they observe, may be formed some idea of the intensity of the fire, whereby an immense mountain has been so far consumed that all the rocks which compose it will crumble into ashes, if the volcano that has produced such an effect should again for a while resume its operations. Contrary to their expectations, they continued to ascend without meeting with any obstacle, passing over a continued series of sloping terraces, of which they reckoned seven before they reached the summit. The sides of the hill they found from top to bottom deeply scarred with ravines formed originally by the torrents of lava, but now serving as beds for the winter cataracts. Among other curious minerals that they met with on their way, they gathered some that they considered as decisive of the fact of Hecla having occasionally thrown out water * as well as fire; and they

* The discharge of water from volcanoes, as well as fire, is by no means unusual. Sir William Hamilton, who most ingeniously endeavors to account for some of the most striking appearances of the globe from this circumstance, considers the water as merely rain that has been deposited in the caverns, contrary, as he says, to the generally received opinion that it arises from a connection between the mountains and the sea. He men-

are from this led to notice an extraordinary matter, of which they do not appear themselves to have seen any symptoms, that so great a quantity of salt * has been found

tions (*Campi Phlegræi, p. 27)* on this subject, that "it is well attested, that in the great eruption of Vesuvius, A. D. 1631, several towns, among which were Portici and Torre del Greco, were destroyed by a torrent of boiling water having burst out of the mountain with the lava, by which thousands of lives were lost."

* This, as they say, (tom. iii. p. 35.) " ne contribue pas peu à confirmer l' opinion de la connexion probable entre la mer et les volcans, tant de ceux qui vomissent des matières embraseés, que de ceux qui vomissent de l'eau alternativement. On peut raisonnablement présumer ces communications entre la mer, les volcans, et les glaciers de la partie orientale, en raison de leur proximité de la mer et la profondeur de leurs racines; ces montagnes vomissent en effet une bien plus grande quantité d'eau que la fonte des glaces ne pourrait produire, et on a même remarqué un goût salin á leurs eaux. On objectera peut-être, á l'égard du mont Hecla, qu'il peut se trouver dans ses entrailles quantité de sel de roche; mais ses entrailles vont jusqu'au niveau de la mer; d'ailleurs indépendamment de l'opinion généralement accréditée de tant de savans de tous les pays, de la connexion secrète qu'il y a entre l'Etna en Sicile et l'Hecla, puisque ces deux volcans ont si souvent brûlés en même temps, on verra nombre d'exemples curieux qui prouvent la sympathie qu'il y a entre l'Hecla, lors

after its eruptions, as has been sufficient to load a number of horses. On the night of the 19th of June, they at length approached the summit, and found themselves on the edge of the crater, in a place covered with ice and snow; yet not of such a nature as that of the glaciers, since it generally melts away in the summer months, excepting only what

de ses éruptions, et les autres volcans de l'Islande plus éloignés de lui qu'il ne l'est de la mer, et même les plus éloignés."—What might be considered as still farther proving the connection between volcanoes and the ocean is, that Ætna is related by Seneca in his second book *Naturalium Quæstionum* to have thrown out a quantity of burning sand; so that "involutus est dies pulvere, populosque subita nox terruit;" but probably that philosopher meant nothing more by sand than minute particles of pulverized matter, a quantity of which, resembling gunpowder, was lately shewn me by the Countess of Gosford, picked up during the last eruption of the same mountain (March, 1809), in the very streets of Messina, fifty miles distant in a straight line, where it fell in such quantities that several cart-loads might have been collected.—The most extraordinary proof of the connection between volcanoes and subterraneous waters seems to be afforded by Humboldt, who, in the zoological part of his travels, speaks of the volcanoes of Quito casting out innumerable quantities of a species of fish that is found in the streams that run into the sides of the mountains.

lies in the hollows and clefts; for Hecla is to be classed among the Icelandic mountains of inferior height, rising to no greater elevation than five thousand feet above the level of the sea. What rendered their walk more uncomfortable was that a flight of snow had recently fallen, the depth of which was not less than a foot and half. Through this they had a long and toilsome passage, before they at last found themselves arrived at the object of their journey, the summit of Mount Hecla*, where the most dreary solitude and silence the

* Sir Joseph Banks thus describes his ascent of the same hill: "we ascended Mount Hecla with the wind blowing against us so violently that we could with difficulty proceed. The frost too was lying upon the ground, and the cold extremely severe. We ourselves were covered with ice in such a manner that our clothes resembled buckram. On reaching the summit of the first peak, we here and there remarked places were the snow had been melted, and a little heat was arising from them, and it was by one of these that we rested to observe the barometer, which was 24. 838. Th. 27. The water we had with us was all frozen. Doctor Lind filled his wind-machine with warm water: it rose to 1..6 and then froze into spiculæ, so that we could not make observations any longer. We thought we had arrived at the highest peak, but soon saw one above us, towards which we hastened. Doctor Solander remained with an

most profound reigned all around them, and they could discover no traces either of fissures in the rock or falls of water, and still less of hot springs or smoke or fire. Though now midnight, it was as bright as day, so that they enjoyed an immensely extensive prospect; looking over all the glaciers to the east, beyond which in the distance towered, like a great castle, the ancient volcano, of Hoerdabreid; while to the north they had a view of all the lofty hills of that quarter, and of a number of lakes of which they could not learn the names. Finding nothing on the top of

Icelander in the intermediate valley; the rest of us continued our route to the summit of the peak, which we found intensely cold; but on the highest point was a spot of three yards in breadth, whence there proceeded so much heat and steam that we could not bear to sit down upon it.—H. 9. . 25. Bar. 24,722. Th. 38. The last eruption of 1766 broke out on a sudden attended by an earthquake. A south wind carried a quantity of ashes to Holum, a distance of an hundred and eighty miles! Horses were so alarmed as to run about till they dropped down through fatigue, and the people who lived near the mountain lost their cattle, which were either choked with ashes or starved before they could be removed to grass. Some lingered for a year, and on being opened their stomachs were found to be full of ashes."—*Sir Joseph Banks' MS. Journal.*

Hecla to induce them to prolong their stay, they descended on the west side by a deep ravine, which, commencing at the summit of the mountain and continuing to its very base, appears evidently to have been the bed of a current of lava, and was most probably formed at the time of the eruption of 1300; since the annals of the country relate that at that period Hecla was rent from the top to the bottom. This cavity has now only the appearance of a deep valley, but it is nevertheless certain, they say, that it was orignally open to the very centre of the volcano, but was choked up by the falling in of stones and rocks, which forced their way downwards on the cessation of the eruption, when the subterraneous fires ceased to lend the earth unnatural strength. Many large masses of rock thrown from the volcano still hang upon the edges of the ravine, where they were cast by the eruption; but far greater heaps of melted and burnt substances are met with at the bottom of this singular and immense chasm.—Thus much for the general and exterior conformation of Hecla. The effects of its subterraneous fires, mischievous as they have been, are small compared to those of other mountains; for which reason I shall proceed to a short description of two or three that have been the most remarkable in this respect.

Krabla, in the north-eastern part of the island, vomited forth great rivers of burning and melted matter between the years 1724 and 1730, one of which was four miles and a half in width and nine in length; it flowed into the adjoining lake, Myvatn, where it continued to burn like oil for many days, filling the lake, drying up its waters, and destroying the whole of the fish. Another torrent overflowed the presbytery of Reykelid, which it so completely buried as not to have left a vestige of the place. These floods of fire are called by the natives *Stenaa* (stone-flood), and afforded, during the day, a blue flame, resembling that of sulphur; but the smoke, which arose from all parts, in a great measure hindered it from being seen. During the night the whole extent of the horizon was illuminated, and the higher regions of the atmosphere became red. Balls of fire were hurled from the stenaa as well as from the burning mountain, and were the means, together with the surrounding redness of the atmosphere, of announcing to the inhabitants at a considerable distance the dreadful disaster.

Katlegiaa in the southern part of the island ejected a great torrent of water on the 17th of October, 1755, which inundated an extent of country fifteen miles long and twelve wide, sweeping away in its impetuous course numerous masses of ice, to which were attached pieces of rock of great size. Whilst the minds of the natives were occupied in the reflection of the dreadful consequences that were likely to ensue from this extraordinary phænomenon, as well as from the accompanying earthquake, a noise was heard like that of thunder, when immediately a rapid discharge of fire and water took place alternately from the mountain, attended by most frightful and horrible roarings, which continued, with but slight intermission, during the whole of the first day; at night the neighboring districts were illuminated by flames, and balls of fire were cast to a great height in the air, so that heaven and earth seemed to be equally in a state of conflagration. On the 19th the column of smoke appeared black in the day, but filled with balls and sparks of fire, which in the night cast a strong light over the whole of Myrdal Syssel, whilst the country situated to the eastward of that district was in darkness both day and night. All the syssels in that direction were covered with black sand and cinders, and loud subterraneous noises were heard even as far as Guldbringue and Kiosar Syssels (eighty or ninety miles distant) and ashes fell like rain in the Ferroe Islands, a distance of three hundred miles!

But the most dreadful volcanic eruption, which the annals of Iceland have yet recorded, took place so late as the year 1783. This was in the south-eastern part of the island, in the district, called Skaptefield's Syssel, and so tremendous was it, that I have been induced here to publish a translation of a very sensible pamphlet respecting it, printed by the Etatsroed Stephensen, who was an eye witness of the calamity; feeling that such an event ought to be recorded in the British language, and being persuaded that my readers will be obliged to me for here furnishing them with it. Without further apology, therefore, or preface, I proceed to say, that the original of the following *Account of the Volcanic Eruption in Skaptefield's Syssel* was published at Copenhagen, in the Danish language, in the year 1785, under the title of *Kort Beskrivelse over den nye Vulcans Ildsprudning i Vester-Skaptefield's Syssel paa Island i aaret*, 1783.* Its author, Mr. Magnus Stephensen, the present Etatsroed of Iceland, upon the intelligence of the eruption reaching Denmark, where he then was, received instructions from the king to proceed to Iceland, in company with Mr. Hans Christopher Diderich Victor de Levetzen, for the pur-

* I have, in the first edition of my Tour (p. 408), been led into an error in consequence of Mr. Pennant's stating, that his account of the eruption of Skaptefield's Syssel was translated from Mr. Stephensen's pamphlet, (See *Introduction to Artic Zoology*, p. cccxxi.), whereas I have since been informed that the original was the performance of S. M. Holme, upon the same subject. The title of the book is, *Om Jordbranden paa island i aaret*, 1783. It was published in Copenhagen, in 1784, and is noticed in a manner not very creditable in the course of Mr. Stephensen's account.

pose of seeing that such measures were put into effect as were deemed most expedient for the safety of the sufferers. Mr. Stephensen was besides more particularly charged with a commission to collect all possible information respecting the eruption, the phænomena with which it was attended, and its effects; to survey, himself, the various currents of lava, and, if possible, the source of the volcano, as well as to search for any mineral substances which were likely to be useful in the arts. Each of the gentlemen above mentioned, on his return to Copenhagen, delivered in an account of the observations made during the journies, which were submitted to his Danish Majesty, who was pleased not only to pay the whole expences of their tour, and allow them a considerable gratuity, but farther to grant to Mr. Stephensen the copyright of his publication.

Such is the substance of the author's *Address to the Reader*, which forms a sort of introduction to the work. The remaining part of the publication I have had translated with all possible fidelity (in part through the kindness of Mr. Jorgensen), and have given it, as nearly as possible, according to the literal sense and meaning of the author.

§ I.

The extent of the damage

"Although no volcanic eruption in Iceland was ever attended with more lamentable consequences than that which took place in the year 1783, yet its immediate effects were not greater or more destructive than many of the former ones. For example, let us only consider what happened in the year 1300, and let us reflect on the long chain of events of which our annals give an account, during the whole of the 14th century, especially in the years 1341, 1350, 1357, 1360, and 1390 *, and on the damage sustained in

* See different annals in *Legati Magnæani Bibl.* in the Royal Observatory, especially in No. 246, among the folios, and No. 407, 411, 418, 421, 425, 427, and 428, among the quartos. See also *Annales Islandorum reg.* among *Langebock's Scriptores rerum Danic, medii ævi. Tom.* III. *p.* 134 and 135.

one morning, when Hecla burned in Bishop Gottsvin's time, about 1436 *. If these be compared with the mischief done in 1783, the difference will appear quite evident: yet it is seriously to be lamented that the damage should reach one of the finest and most beautiful parts of the country, formerly supporting a number of people, who are now reduced to a state of misery and ruin. However, praise be to God, the injury is not nearly so extensive as many erroneous reports have led people to believe.

§ II.

The state of the weather in the spring of 1783.

A delightful spring succeeded an unusually mild winter in Iceland, in the year, 1783. Clear, calm, and warm weather, with sunshine, were only interrupted by soft breezes from the south, mingled with abundant showers of rain. The pastures were at an early period seen dressed in a green and luxuriant vegetation, and, in the month of May, adorned with many herbs and flowers in their freshest

* See No. 213 folio, and 407 quarto, in *Leg. Magn. Bibl.*

vigour. The greatest benefit was anticipated from the cattle, which had become sleek and strong after so mild a winter and spring, and every one rejoiced at the prospect of a fruitful summer and an abundant harvest. But all these happy illusions fled with the month of May.

§ III.

The commencement of the Eruption.

Towards the latter end of May a bluish and light smoke, or fog, was seen floating along the surface of the earth, and attracted the notice of several well-informed people: yet no one had the smallest idea of the approaching evil till strong earthquakes were perceived and felt over the whole of Skaptefield's Syssel on the 1st of June. These became daily more terrible, especially during the mornings and evenings, and at last on the 8th of June, the first day of Whitsuntide, they announced the most violent commotions in the bowels of the earth.

At eight in the morning the weather was still fine and clear, but towards nine a dark and black bank of smoke arose in the north, and at length extended itself over the district called Sida. This bank could not at first be seen from the farm-houses that were pleasantly situated at the foot of a lofty and closely-connected range of mountains, stretching for the most part from east to west, nor could the inhabitants distinguish it till it was quite near, and immediately over Sida, but several persons who were proceeding from the district Landbrot, situated a little to the south of the place just mentioned, to Kirkebai-cloister church, observed a great number of pillars of smoke arise from among the wild mountains in the north, and in a little time gather themselves together and form the large black bank. On the nearer approach of this, Sida became involved in darkness, and, when the bank was perpendicularly over it, an immense quantity of sand and ashes, much resembling those of burned coals, fell upon the ground, covering it to the thickness of an inch. Intermixed with these substances was one of a grey, shining, and hard nature, which will hereafter be more fully described. A southern wind prevented the farther progress of the bank on this and the following day, but earthquakes, with heavy peals of thunder, together with subterraneous noises and cracklings, continually increased; so that during the whole day, and long after the close of it, such noises were heard as might be compared to the roaring of a number of cataracts all meeting in the same place, or something similar to a large kettle boiling over impetuously.

On the 10th of June several fire-spouts were distinctly seen, for the first time, rising from among the mountains towards the north. The black bank became more lofty every day, while earthquakes, peals of thunder, and strange sounds increased.

§ IV.

The river Skaptaa disappears.

The Skaptaa was formerly a very large river that flowed between Skaptartunga and Sida, and for the most part ran parallel with the latter, which was also divided by this river from the districts Landbrot, Medalland, and Skaptartunga. It took its rise from Sida or Skaptar-Jökelen, an ice-mountain, situated about nine leagues north of Sida. The stream was very rapid, and the river in many places passable only in boats. In the spring of 1783, a vast quantity of fetid water, mixed with gravel or dust, was observed running down the Skaptaa, which was greatly swollen on the 9th and 10th of June, when, to the astonishment of every one, it totally disappeared, and was so dried up in less than twenty-four hours, that people walked across with ease in such places as were formerly crossed by travellers with difficulty in boats. There is, however, still to be observed a small running stream; but it only arises from a number of little brooks which, proceeding from the sides of the mountains, discharge their waters into the bed of the Skaptaa. These waters, in Iceland called Berg-vand (to distinguish them from the thick and milky Jökel-vand, of which the Skaptaa and all other rivers deriving their source from ice-mountains consist), were quite clear and pellucid. So remarkable a phænomenon as the drying up of the Skaptaa, was fully accounted for on the ensuing day, the 12th of June, when a dreadful fire-stream came pouring down with the greatest impetuosity, like a foaming sea, into the

Skaptaa. This river every where ran through deep vallies and between lofty cliffs, which were in many parts from four to five hundred or even six hundred feet high; yet the fire-stream not only filled up these cavities, but actually overflowed a considerable tract of land on both sides. It is only in a few places that there are still to be seen above the lava some of the tops of the highest mountains that formerly enclosed the Skaptaa.

§ V.

The state of the atmosphere on the 12th of June.

It is impossible to find language that will convey an adequate idea of the horrible circumstances that accompanied the first eruption, and made this day peculiarly dreadful.—A dark and dismal bank arising in the north-west and pouring forth ashes, sand, brimstone, and the hard greyish substance before alluded to.—An intolerably stinking and suffocating smoke, concealing the face of the sun and absorbing its brilliant and beneficent rays.—Seldom could this luminary be perceived through the thick and sulphureous steam, and when it now and then became visible it had the appearance of a globe of a gloomy and blood-red color.—Constant shocks of earthquakes, innumerable fire-spouts in the north, a dreadful foaming stream of fire rushing down into the Skaptaa, indescribable sounds in the air, heavy subterraneous thunders, noises from the mountains and continued lightnings, filled every breast with the greatest terror, and led the poor inhabitants to expect every moment that heaven and earth would be annihilated. Nor is this to be wondered at; for none of the wretched people could tell how soon they and their property might become the prey to such powerful and visible means of destruction.

§ VI.

The progress of the fire.

The stream of fire, though now running with incredible fury, for the most part, along the channel of the Skaptaa, nevertheless, here and there extended itself over the old lava-tracts on the sides of the river. Great cracklings and noises were heard, when many pieces of red-hot lava fell together into holes in the rivers, where the water had been previously evaporated by the fire. The current of lava had in a single day, before the evening of the 12th, proceeded as far as the farm Aa, in Sida, where it overflowed houses, enclosures, pasture-lands, and carried every thing away before it. It had also in another direction done much damage to the farm Buland, and destroyed Svartanup and Litlanes. On the western side the fire had already extended itself as far as the farms Svinadal and Hvaam, where much injury had been sustained. The same was the case with Skaptardal, on the eastern side.

According to all appearances it might reasonably have been expected that the immense masses of lava, rushing down like melted metal from out the Skaptaa, with such prodigious force and velocity, would at once have over-run Medalland, which lay just beneath, and consequently have done infinite mischief; but at this very place the fire was arrested in its progress, on the succeeding day. A lake, formerly situated in a place between Skaptardal and Aa, but now in part filled up with sand from the Skaptaa, swallowed up a vast quantity of lava that, for several successive days, poured down with a horrid noise. The fire-stream was consequently very much diminished, but when the great lake was at length filled, and when the lava, by continual supplies from its principal source, had risen to a considerable height in the valley between Skaptartunga and Aa, then the stream extended itself much farther over the lowlands. Frightful noises and sounds that caused the whole place to tremble, strong claps of thunder and constant lightnings, prevented the inhabitants from taking any rest between the 14th and 15th of June. The burning lava was seen at that time to overflow the farm Nes, in Skaptartunga, together with the whole of the adjacent country, and, among other places, several that were well wooded between this farm and Skalarstapi, in Sida. Another arm proceeded eastward from this place, passing by the farms Skal and Holt, where it stopped several days; but during that period burned the wood-lands called Brandeland, belonging to Kirkebai-cloister.

On the 13th previous, several persons had endeavored to go up into the mountains in order to discover the real source of the fire, and the extent of the mischief that had ensued from it in the district; but the thick smoke issuing from the lava made their attempt quite impracticable. Nothing could be seen but the stream that had filled up the Skaptaa, together with innumerable fire-spouts, which rose out of the river, close by Ulfarsdal and a long way towards the north.

§ VII.

The fire-stream. During the ensuing three days, till the 18th of June, the fire spread itself slowly towards the south and south-west, from Skal over the old lava-tracts. It penetrated the innermost and most concealed crevices, by which means the old lava was as it were lifted up from its original bed; and formed into a number of hills. It did not, however, suffer itself to be so removed without a strange whistling kind of noise, caused by the fire forcing the air from the subterraneous caves through the cracks and narrow openings. Sand and earth were only slightly scorched by the fire, and it had but little effect upon the grey-stone *(graasteen);* but, on the contrary, it penetrated into the smallest fissures, and pores of the old lava, which was soon melted, flowing with the new, and often taking fire itself. Thus, when an old piece of lava was melted, it immediately lighted that with which it was in contact, and so continued till the whole was on fire. It is remarkable that, during the melting of this lava, the uppermost crust remained in its original state, so that large pieces might be taken out of the fire-stream which had the appearance of beaten metal. The stream forced itself downward, where it continued its progress, throwing the above-mentioned crust up into the air or to the sides of the current, in which situation it remained for some time, burning in a pure steady flame. Wherever the fire-stream had in this manner forced its way under hills and rocks, they were, by means of the heated subterraneous vapors, thrown into the air with prodigious force and a dreadful noise. It may easily be conceived what a horrible crackling must attend the bursting asunder of such immense masses of rock, many of them from an hundred and twenty to an hundred and eighty feet in height; but how much more terrifying must be the falling of these bodies, when the velocity is so much accelerated by the vast height to which they are generally thrown. During these days the fire increased so much in redness, and spread itself in so great a degree over Sidumanna Afrett*, towards the south, on account of a northerly wind then blowing, that several farmers residing on the heaths actually fled with all their cattle and moveables, frightened at the immense conflagration, which, though at a considerable distance from them, appeared to be quite in the vicinity.

§ VIII.

New eruption of lava.

On the 18th of June, a most dreadful eruption of lava again broke out among the mountains. In those places where the Skaptaa had not been quite filled up, the lava was now observed to rise to a height, far exceeding that of the steep-

* *Afretur,* in Icelandic, signifies a wild and uninhabited tract of mountains, covered with grass, where sheep and cattle are sent to fatten in the summer. The tract here alluded to belonged to the *Sidumen,* or Men of *Sida,* that is, people who resided on *Sida,* and therefore it is called *Sidumanna afrettur.*

est mountains that enclosed the river, and to rush forward over such tracts as had previously been destroyed, and even so far as the utmost extremities of the current which had cooled and become stationary.

In the middle of the lava were to be seen red-hot rocks, which the stream had torn from their beds. A thick, white, and suffocating steam issued from the two rivers that had been intercepted by the fire, and were constantly boiling; and the vast quantity of hot water, which overflowed the meadows and pasture-lands, did no inconsiderable damage, especially near the farms Svinadal and Hvaam, in Skaptunge, as well as in the eastern quarter, where a tract of woodlands belonging to Skaptardal was totally destroyed.

§ IX.

Progress of the fire-stream.

On the 19th of June the fire extended much farther, dividing itself into two branches; the one rushing on, with the same rapidity as it did the preceding day, in a southern direction along the river Melquiol and over Medalland: the other moving towards the east and along Sida, where it burned the country about Skalarstapa, and forced itself with incredible fury up to Skalarfiall; but, as this mountain checked the more rapid progress of the fire towards the north, the lava rose considerably in height, and, in ascending the sides of the mountain, rolled up its covering of moss in the same manner as a large piece of cloth might be done by human means. In the evening the stream was not above an hundred and twenty yards distant from the church of Skal, when the inhabitants quitted it. They had been in hopes that the fire would have spared this place, as it had passed by, four days preceding, without doing any injury; but just at this moment, contrary to the expectation of every one, the fire broke out afresh, and carried away every thing that had before been left by the lava. This eruption was accompanied by a strong and constant trembling of the earth, which had much abated since the 12th day of June, the first of the eruption. On the ensuing day, the fire-stream proceeded to the farm Holt, overflowing the tract between that place and Skal, by which means the lava that had lately reached that spot, rose considerably in height. The other branch, previously mentioned as having bent its course towards the south, along the river Melquiol, extended itself widely on both sides of it.

§ X.

Destructive consequences.

Notwithstanding that the farm of Skal was placed in an elevated situation, at the farthest extremity of the great valley, or Skaal, whence it takes it name (*Skaal* in Icelandic signifying a *bowl*), yet the lava had prevented all access to it; and when, on the ensuing day, the 21st of June, great torrents of rain had swollen the brooks, proceeding from the mountains on both sides of the farm of Skal, this place was, together with the church and adjacent houses, entirely overflowed with water, which the next morning was boiling excessively. At the same time that the flood destroyed Skal, the western branch of the fire-stream spread itself with great rapidity farther to the eastward, over the river Steinsmyrarfliot, and all the way to the parsonage, Holmasel, which, as well as the church, houses, and the whole neighboring country, were, on the following night, entirely covered with lava. The farm of Holmar, likewise, shared the same fate.

§ XI.

Farther account of the damage sustained.

On the day ensuing, the 22nd of June, the fire continued in its progress along the river Steinsmyrarfliot, and close to the farm Efristeinsmyry, where the lands were much damaged; but here it changed its course, proceeding towards the south from this place, passing the farm Sydri-Steinsmyri, which consists of five separate buildings, and stopping about eighteen hundred feet from the most northern of these, where, however, no considerable injury was sustained.

The fire-stream spread itself greatly towards the west, over the river Fedgaqvisl, and overflowed the farms Sydri and Efrifliota, together with the houses and lands. Although the farm Hnausa has not been destroyed by the fire, yet the rivers Steinsmyrarfliot and Fedgaqvisl, in consequence of their being dammed up, had caused it to be almost buried under water, which finally proceeded along the channel of a small brook, that used to run close by the house, and is now quite impassable.

The lava farther continued to overflow the farm Botnar, and much pasture-land, as well as the country between the river Landa (which had been filled up with lava), and the farm Hnausa.

§ XII.

New eruptions. From the 22nd of June to the 13th of July fresh streams were observed to proceed along the Skaptaa, and extend over the lowlands. Between Skalarstapi and Skaptartunga the lava had risen into a lofty hill, from the continual eruptions, and had become, towards its extremity, firm and solid, which prevented the new streams, that were pouring down the mountains, from having a free passage, causing them to divide into various branches; of these, two flowed along the western and two along the eastern side. One of the former of these passed over Neshraun and the farm Hnaus, of which place nothing is now to be seen, except a small sheep-cote. The other western branch, which proceeded along the river Landaa, overflowed the farm Nes, together with the houses, fields, woods, and meadows belonging to the parsonage Asar, as well as most of the lands belonging to the farm Ytri-asar. The priest saved the greater part of his effects that could be removed, and afterwards set off for the western part of the Syssel.

From these farms the fire-stream over-ran the southern district beneath, advancing towards the west, along the broad channel of the Kudafliöt, one of the largest and most remarkable rivers in the country. It stopped, however, a little to the north north-west of the farm Leidvöll. To the north of this place a great bight is observed running into the lava-tracts, of which, indeed, only one point has reached the Kudafliöt: the rest having passed the most eastern extremity of this farm, bending more and more in that direction to the north of Stadarholt; thence again, in the same course, immediately to the north of Hnausar.

In the mean while, one of the eastern branches before mentioned proceeded over the Landbrot, along the Skaptaa, which led to it, and destroyed in the way several places; stopping at last in the midst of Hraunsmelar, in Landbrot. The other branch ran along the Sida mountains, overflowing, on the 2nd of July, the church and all the houses of Skal which had previously been deluged with water, as well as all the lands and houses belonging to the farm Hollt, together with the excellent meadows that lay to the south of it. It dammed up a small river that ran close by the farm, and on the 6th of July, buried Hollt itself in the lava.

Hence the fire proceeded eastward, and between the 14th and 17th of July continued its course along the river Skaptaa, over the river Fiadra, which was quite choked up by it. Nevertheless, the greatest quantity of lava flowed over the lofty waterfall Stapafoss, in the Skaptaa, and at last filled up the enormous cavity which had for so many succeeding centuries been hollowed out by the waters of this great cataract. Near this place the lava over-ran the farm Dalbai, in Landbrot, with all the houses, and the greater part of the meadows and pasture-lands, after having done much mischief to the farms Heidi and Hunkurbacka, upon Sida. These, however, are not so much damaged, but that each is still capable of supporting a family. The farm Holmar, in Landbrot, was also somewhat injured, as it was threatened on the north side by the approach of the lava-stream, and on the opposite one by the water which had been impeded in its progress by the fire. At length, on the 20th of July, the fire ceased immediately west of an insulated rock, called Systrastapi *, which lies, at the utmost, one mile west of Kirkebai-cloister.

§ XIII.

Another fire-stream east of the former.

Hitherto I have confined my account to the most material injuries that have been occasioned by the great western fire-stream, or that which took its course along the Skaptaa, and the destruction caused by it in Medalland, and the countries adjoining the Skaptaa, which lie

* *Systrastapi* has received its name from a traditionary story about two sisters, who, it is said, were discovered fighting on this rock.

to the west of the river Hverfisfliot. In what I am now going to relate, it will be seen that the fire, to the eastward, has raged with a fury equal to that of the great western stream, and exhibits a spectacle equally melancholy and distressing.

The first scene was disclosed on the 28th of June, when a thick and black bank of sand and smoke, proceeding from the place of eruption, and driven by a strong breeze from the north north-west, towards Fliotshversi, caused such a frightful state of darkness over the whole of that district, as well as over the eastern part of Sida, that, even at noon, it was impossible in the houses to distinguish a sheet of white paper from the black walls. On the 14th previous, indeed, a degree of obscurity, equally uncommon, but not so terrific, was experienced in the middle of the day thoughout Sida: but it was only during the present interval of darkness that a number of red-hot flat stones, with enormous quantities of sand and ashes, which entirely burned up the grass in the pastures, fell upon the whole of Fliotshverfet, the two farms Nupstad and Raudabag only excepted.

These substances poisoned the earth and water, rendering them alike destructive both to man and cattle, and threatened to set fire to the houses themselves, whenever any of the stones and ashes happened to fall upon them.

§ XIV.

Hverfisfliot disappears.

On the 3rd of August a great smoke was, for the first time, observed to arise from the Hverfisfliot, and the water was found to be excessively hot. This river, it may here be proper to remark, was equal in size to the Skaptaa, but infinitely more dangerous to travellers, in consequence of the rapidity of the current and the great insecurity of the ground. Its heat continued daily to increase, till, at the expiration of a few days, the waters were entirely dried up. This circumstance filled the inhabitants of the district with the greatest fear and consternation, who, already terrified at the mischief that had been sustained by their neighbors, after the drying up of the Skaptaa, anticipated similar misfortunes, on observing the disappearance of the Hverfisfliot. The result convinced them that their fears were well grounded, and proved to them, that in this instance, also, the same phænomenon produced an equal or even a greater degree of danger.—Dreadful pillars of fire were seen rising at a great distance among the mountains in the north, on the morning of the 9th of August. They appeared to approach nearer each other, and at last to form, as it were, a wall or lofty bank upon the earth. Continual lightnings, with strong hollow sounds, somewhat resembling thunder, were also heard in the same direction. A foaming fire-stream now broke down into the channel of the Hverfisfliot, urging its course with incredible and matchless fury. The stream spread far and wide over the extensive tracts of sand, situated in the south, and in one single evening overflowed more than four miles of ground, in that direction from Orustuhol (a hill so named from duels having been formerly fought there), and entirely blocked up the road between Fliotsverfet and Sida.

Continual eruptions from the mountains increased the extent of the tracts of lava, so that, at the latter end of August, they entirely over-ran the farms Eystradal and Thverardal. All the adjoining houses and the greater part of the enclosures were buried under the lava, so that the places where the buildings formerly stood are no longer visible. These latter eruptions, also, have done much damage to the farms Selialand and Thvera, and the inhabitants were entirely frightened away from the parish of Halfafells, although no houses were destroyed in these places.

The eastern fire-stream broke out at a much later period than the western one, and continued raging much longer, frequently at intervals bursting through the crust or surface, which had become indurated. It is even asserted that in February, 1784, a fresh eruption proceeded from the mountain, and caused the lava, in the eastern branch, to rise to such a considerable height as it now is.

§ XV.

So much for the progress of the fire, and the immediate destruction occasioned by it. Were I here to relate all the contradictory, insignificant, ridiculous, and superstitious accounts that are reported of the eruptions, it would be both an useless and a tedious task, especially since the greater number of such accounts are the offspring of fear and ignorance. But two questions naturally present themselves to our minds, which are too closely connected with the subject to allow me to pass them over in silence.—The first is, "whether the subterraneous fire is to be deemed only an eruption, or the earth itself is to be considered as ignited; or whether these two causes may not have operated together?" I am well aware that the greater part of those persons that were on the spot bring forward various arguments in order to prove that the earth itself was ignited, but the reasons they assign appear to me to be weak and in themselves highly contradictory. It is remarked, that before the liquid lava had over-run several of the places now burnt, fire had actually, here and there, broken through the soil. We are informed that the proprietor of the farm Botnar, in Medalland, had, on the first breaking out of the fire, collected eighty sheep and placed them, as

Respecting the nature of the fire.

he supposed, in a state of security upon a small island, but that, before the farmer had returned to his own house, the fire appeared to break out from that very island, and he had the misfortune to be the sad spectator of the ruin of himself and family. This account, indeed, was, as far as I was able to ascertain, perfectly correct, but nevertheless it does not at all prove that the accident was caused rather by the earth itself being on fire than by a fire-stream: for, at the very moment that the farmer had collected his sheep upon this spot, the lava was rushing along with the greatest imaginable rapidity, and took quite a different course from that which was at first expected; proceeding towards a neighboring river and along its channel, till it arrived at the island, which it burned together with the sheep.

The Icelandic annals relate a long series of such eruptions, continued through whole centuries; but we do not find any account distinctly describing the nature of the lava-streams which formerly over-ran whole districts. The damage sustained is simply noticed, but upon the subject of the progress of the fire, authors are entirely silent. It is therefore quite natural that the late fire should appear particularly frightful to the spectators, and that they should be led to suppose that the fire broke forth from the entire and uninjured crust of the earth, at considerable distances from the fire-stream itself. My own opinion is that such appearances always exist when the stream of lava, for the first time, pours down from the mountains upon a fertile tract of land, or upon a soil that is loose and free from obstructions. We may readily imagine what an immense weight must fall upon the earth, when the lava rushes down from the high to the low lands, and we may in like manner conceive it possible for the lava to burn and force itself a passage to a considerable depth below the surface of the earth. The lava itself being a fluid, driven forward by every new accession of matter, it can, without doubt, proceed in its course as well below as above the surface of the ground, and even in some instances with greater rapidity. Above it must work its way over all inequalities, and where the stream of lava has to cross a valley in its course, its progress is necessarily impeded till the hollow is filled up: it must, too, carry with it a great number of stones and other things, with which such streams of lava are filled. Beneath the surface, on the contrary, where only that lava can penetrate which is in its purest and most fluid state, it finds its way into many places through a more loose and open soil, percolating like water, continuing the same even course, and is not stopped by the above-mentioned inequalities; but flows forward uninterruptedly to a great distance. The fire having in this manner forced its way down into the earth, and proceeded forward, it is easy to conceive that, in consequence of the vast heat arising from so much burning matter, the damp and moisture would be converted into vapors, and that these, by the force of the lava, would be driven up through the crevices, to the surface of the earth, appearing above ground in flames. Where, indeed, the soil is full of rocks, it is not possible for the fire-stream to proceed with any velocity beneath the surface, much less through the old lava-tracts, which are of considerable depth.

From these considerations I conclude that the fire has never been known to proceed, either from the late volcanoes in Iceland, or those in any other part of the world, through the uninjured crust of the earth, at a distance from the lava-stream, except, indeed, from the causes just mentioned.

I really believe that there is no more reason for inferring from the fire bursting through the earth at a distance from the stream of lava, that the earth itself is ignited, than there would be to draw the same conclusion from a simple eruption. And among other forcible arguments, in favor of my opinion, it strikes me that the continual supplies of matter which the streams receive are a strong proof that such a fire is really the effect of an eruption. They imply one common source and one spot whence they originate, therefore properly belong to one eruption, and are by no means caused by the earth itself being on fire: for, had this been the case, I should imagine that burnt matter must have been found at a much greater depth below the surface of the earth than is the case on open plains, where it seldom exceeds six or eight feet, though in other places it is much deeper; for instance, if it has been impeded by any thing in its progress, or if it has accumulated in a valley or river. Nor have I been able, though I took great pains to ascertain the same, to discover, either in the vallies or in the mountains, iron or sulphur, nor, indeed, specimens of any combustible soil different from that which may be found to exist where volcanoes have never operated.

§ XVI.

The above must be considered as an answer to the first question, and I think I have proved that the subterraneous fires, which have broken out in various places, have been the consequence of an eruption and not of the earth itself being on fire: and this answer will naturally lead me to another question, "Where then are we to look for the original source of the eruption?"

Question concerning the place of eruption.

If we were to rely on the many oral assertions, as well as on those that are committed to writing, concerning the fire, we should be led to conclude that its origin was not in one but in many places. At least, according to the generally received opinion, one place must be allowed for the eastern and another for the great western stream of lava; for so did it appear to those persons who in 1783 proceeded to some distance up the mountains. In like manner the tremendous pillars of smoke among the mountains seemed to the inhabitants of the plains to have various sources, and the same also seemed to me to be the case last summer, when I was in Sida. From later observations, however, I am induced to adopt a totally different opinion. According to a part of my instructions I resolved to undertake a journey myself from the plains to the place of eruption, notwithstanding that every one represented the accomplishment of my design as a thing impracticable on account of the great distance, the badness of the roads, the fresh streams, the impassable rivers, the intolerable heat, the dreadful smoke, the suffocating smell of sulphur, the want of grass and forage for our horses. All these, however, could not deter me. With a great deal of difficulty did I at last persuade a brave old man, who had been born in this district, to accompany me to the mountains, at least as far as the place whence it was said that the eastern stream of lava had its source. It now only remained for us to consult from what place such a journey might most conveniently be undertaken. From Fliotsverfet it was impossible on account of the impassable mountains, especially the ice-mountain, Sidu or Skaptar-Jökul; and moreover the new lava-tract passes in that direction, as well as the Hverfisfliot, which is quite choaked up with lava. From Skaptartungen on the western side it was likewise impracticable to proceed, as the Skaptaa and the other two great rivers Efri and Sydri-Ofæra, which had been filled up with lava, impeded the progress in this direction. It was therefore necessary that the journey should be commenced from Sida, and with this view we procured two horses to convey the two boring instruments (an earth and a mountain-shaft), with some provisions and a small tent. To these were added six riding-horses for myself, my companion, and servant, which latter was to assist in making experiments with the shafts as well as to take care of the horses.

§ XVII.

Journey to the mountains.

In the morning of the 16th of July, at four o'clock, I proceeded on my route towards the mountains, and on advancing near the moors, north of Prestbacke and Mördtunga, upon Sida (for my tent was pitched between these two farms), I was greatly astonished at the miserable appearance of the pastures. Even here, where many farmers from Sida were in the habit of sending cattle in summer, for the sake of the excellent grass, every thing was evidently quite burned up by the falling of hot ashes and sand; excepting only in those places where an enormous mass of volcanic ashes and gravel had formed a deep black covering, and thus wholly concealed the surface of the earth. The farther we advanced towards the mountains, the thicker lay the ashes, reaching in some places even to the thickness of four or five inches: yet even here in a few spots some half-withered herbs and blades of grass were beginning to make their appearance. The same fate had attended Sidumanna-afrett formerly so fertile in grass. In this place ashes and sand lay still deeper, and not a single trace of herbage was to be seen; so that it cannot be expected that vegetation will recover itself in a less space of time than four or five years; although there is reason to hope that in two years, provided no new eruption ensues, the other lands may again become in a measure fertile.

After crossing a number of dangerous moors, I arrived at a very large and well-known mountain, called Kallbakur, near the eastern stream of lava. From the north-eastern extremity of this eminence I could at one view survey a vast tract of lava which had proceeded a considerable way towards the west, passing the north side of Kallbakur, between that mountain and another north-west from hence, called Miklafell. Between Kallbakur and Eriksfell, a mountain on the eastern side, the lava became very narrow, nor does it spread itself much on either side of the former channel of the Hverfisfliotet, where the smoke was yet rising to an alarming degree. No accumulation of clouds in the air can be imagined to form a more dense body than the smoke that now issued from this place, which rolled itself over twice or thrice before it could be driven farther on by the wind. Hence I followed the lava-stream, which spread itself a long way towards the west, between Kallbakur and Miklafell, in the north, at length approaching so near to the eastern side of the latter mountain, that it was with great difficulty we could pass between it and the hill; especially as the smoke was here very strong. On the western side indeed there is a road, but it is nearly impassable; so that nothing was to be done but to proceed over the middle of Miklafell. Here I alighted, and, having given the horses a little rest, went with my companion over the lava, as far as the heat would permit us. I examined with all possible diligence the different sorts of lava, and whatever else was remarkable, of which an account will be given in its proper place, as well as of the result of the experiments with the boring instruments.

We now proceeded farther upon our journey, endeavoring to ascend the mountain on the south side; but it was not possible here to continue our route, except on foot, and in this manner I at last, with great difficulty, reached the summit, though with only one of my companions: for the other was obliged to take a circuitous way through a valley, in order to get the horses over. The prospect from this mountain was truly melancholy. Towards the east was seen the new black lava, close by the huge ice-mountain, while the rest of the picture presented to the view nothing but the prodigious quantity of ashes and sand that were immediately after the commencement of the eruption scattered over the whole of Sidumanna-afrett. Following the lava-stream still farther towards the north, by sun-set we arrived at Blæng, a very lofty mountain and the most northerly one here known. With great difficulty did we scramble up, and observe that the stream of lava (which had passed close by the south-east side of Blæng, stretching somewhat to the westward along the south side, where a lake is seen in one of the vallies) advanced more and more to the west on the north side of the mountain, and appeared to form an arm in a south-westerly direction towards the channel of the Skaptaa. The smoke that issued from the lava-stream north of Blæng had an appearance equally terrible and indescribable. It intercepted our view from this place, which was otherwise very convenient for the purpose: yet we could discern a considerable hillock, or small mountain, greater in its diameter than in its height, whence there also proceeded a thick and black smoke. There I concluded must be situated the source of the eruption, and immediately advanced on my journey thither: but I soon found the difficulty of such an undertaking, as I continued along the lava-tract north of Blæng. The hazard was increased by the extreme brittleness of the pieces of lava and the impossibility of finding a secure footing; and when we advanced about sixty or eighty yards upon the lava it became more and more dangerous and insecure, and at the same time burning hot, so that it was no longer possible to stand upon it. The smoke, too, that rose surrounded us on every side to that degree that we scarcely knew on which side to turn in order to retrace our steps. At length, however, we effected this, and I attempted again to pass the lava in another place east from Blæng, in hopes that the mouth of the volcano might be approached on the north side; but here was experienced the same strong and insufferable heat as at the former place, so that I was obliged to return the next morning at sun-rise, after having employed the whole night in vain, in endeavoring to get over. I still persisted in advancing along the lava for a considerable distance by the west and south-west parts of Uxatindur (where the lava-stream was very narrow) in order to cross, but my labor was all fruitless. The heat was intolerable, and when I began to make use of the boring instrument, it became, at the depth of four feet from the surface of the earth, so hot that it was with difficulty we could draw it up again, though our hands were protected with mittens. When I found that my people could no longer bear to work with the mountain-shaft, and that the great heat was likely to render our experiments useless, we moved on towards the outskirts of the lava, where the temperature was more supportable, and there continued our observations.

§ XVIII.

Source of the eruption.

It is I think certain, that the place whence the eruption had its origin, is that small low mountain, which I have just mentioned as being situated to the north of Blæng, and which is, indeed, the most northerly one we could discover. That its source is not farther in that direction, we have the strongest proof in the Skaptaa; for, had such been the case, this river would have been filled up long before, in which case it must, wholly or in part, have made its way along the east side of the lava-tract, north-east from Blæng, instead of taking the western, as at present. In like manner, could the stream of lava have flowed farther towards the north, in that extensive tract of country between those spots, where the sources of the two great rivers, Skaptaa and Hverfisfliot, seem to lose themselves in it, the rivers must have been sooner choked, and there would consequently have been a great deal of stagnant water, which is not the case at present. It is an undoubted fact, that the whole tremendous current of fire has proceeded from one common source, and in all probability this source lay in the small mountain I have been speaking of; for, from that place, the whole range of country overflowed with lava gradually slopes as far as the Skaptaa, which is also the case all the way from Skaptaa to the sea. Whether this place of eruption was originally a mountain, or whether it has been formed by the ejected matter, cooling around the crater, into its present form, I cannot determine. But thus much is certain, that never since the island has been inhabited has there been known in this spot an eruption or a volcano, not even when Katlegiaa, which is situated a little to the westward, or when several other mountains to the eastward, have raged in their greatest fury. Still, however, there are observable, in the northern part of Sidumanna-afrett, in the Borgden, and in the Landbrotet itself, evident marks of subterraneous fire, which must have been in an active state, at a very distant period of time, since our annals are entirely silent upon the subject.

§ XIX.

Continuation of the journey through the mountains.

Great as was my anxiety to approach near the seat of the fire, yet the impossibility of doing so rendered all farther efforts vain, and I was obliged to return. I felt also considerable apprehension for my horses, with which I had now travelled for three successive days, through long and dangerous roads, without their having the least forage or grass; and it may naturally be conceived, that both myself and my companions were very much exhausted by climbing day and night, by ascending and descending the heated lava, and by boring with the mountain-shaft, whilst enveloped in a thick sulphureous stench. On my way back I continually followed the course of the lava, commencing at a place a little way distant from Uxatindur (where the last attempts were made to use the mountain-shaft, and also to cross the current) as far as Helisa, a river which formerly fell into the channel of the Skaptaa. During this journey the dreadful smoke arising from Hrossatuna, presented a remarkable appearance, as far as the lava had filled its channel. It stood here like a thick wall, forming a direct line with that of the stream. Yet still the smoke was not so dense in this situation, as that was which arose towards Skarptargluifret, for this latter was at a very great distance distinguishable from the other, and we were consequently enabled to trace the exact course of the whole channel of the Skaptaa to the northward, as far as the torrent of fire had filled and overflowed it. It was impossible to think of crossing the stream of lava at this place; for even on the plain we had found it so excessively hot, that on three preceding trials, we had been under the necessity of returning with the greatest rapidity, to avoid being burnt; and, if such were the case in a level spot, how much greater must the heat have been in the river itself, where the lava lay, without any exaggeration, from four to five hundred feet deep. The quantity of smoke, too, was a certain mark of the vast heat of the mass before us; and to the difficulty of proceeding in consequence of this impediment, was to be added another, arising from the waters which had accumulated, both in the lava itself, and on its western side, by the stoppage of the great rivers Skaptaa, Sydri, Efri-Ofæra, and

Hrossatuna, together with several lesser streams. The eastern branch or arm of the lava was precisely in the same condition; the river Hverfisfliot being there stopped and entirely choked up.

But as neither the exhausted state of our own strength, nor that of our wearied and famished horses, would allow of our taking an exact survey of the whole western extent of the valley, I changed my route in the afternoon, and turned off from Hellisaa, taking the shortest road to the eastward past Geirlandshraun; and, at length, at one o'clock in the morning of the 18th of July, regained my tent in safety, between Prestbakke and Mordtunga, on the Siden.

§ XX.

Length and breadth of the lava.

The extent of country destroyed, together with the progress of the fire, is thus, as I hope, sufficiently and circumstantially described. It will here be proper to remark that the western current of lava in its longest or narrow arm out of the valley, does not exceed six Danish (or about twenty-four English) miles, in length, even including all its windings as far as Svartenup. From this place it stretches over the valley towards the south, to a distance not exceeding three miles and three quarters, so that its greatest length, taken from the volcano itself, cannot be reckoned at more than nine miles and three quarters, or, at the utmost, at ten Danish miles. From the farm-house of Skal, which, together with the church and other buildings, were deluged and covered by the torrent, it runs two miles and a quarter towards the south, and in general cannot be estimated at more than two miles of breadth, in the Medalland, where it has nevertheless most extended its ravages, and has done the greatest damage.

From what has been now said, it will readily be perceived that the actual destruction, caused by the fire in the district of West-Skaptefield, is by no means of the extent that many people have described it to be; and, as I have always regarded it to be my sacred duty to adhere to the truth of facts, as far as it has been in my power to ascertain them, it is impossible that my account of the eruption should coincide with that given in a publication by the student, Mr. Sæmund Magnussen Holm. A work so inferior and faulty in its nature, does but little honor to Danish literature, and still less to its author; yet it is now not only widely circulated at home, but has likewise abroad been honored by the decoration of a foreign dress, and may possibly be received, and readily credited, among such persons as have not had the opportunity of obtaining more correct knowledge. With regard to Mr. Holm's account of the fire, after having myself personally investigated the spot, and correctly estimated the extent of the damage sustained, I dare venture publicly to assert that his description is faulty to as great a degree, as is the difference that will be found to exist between his two geographical charts and that which accompanies my statement.

It is a matter of real satisfaction, that the estates of Holmur, Hunkurbackur, Heidi, Skaptardalur, Hvammur, Svinadalur, and the farm of Buland, together with the church at that place (which Mr. Holm describes as being totally destroyed by the stream of fire), are all yet standing in good condition, not having by any means sustained so much damage as has been stated by him. And it is also a subject of rejoicing that, instead of eighty-nine farm-houses which he informs us are all damaged, and the greater part of them destroyed by fire and by large pumice-stones, neither Mr. Livetzen, nor myself, could in the course of our journey (which extended to every farm in the way of being injured by the eruption) discover more than eight farmers'-houses, and two cottages, which can never be inhabited: these are, Eystridalur, Thverardalur, Aa, Nes, Hölmasel, Holmar, Efri-Fliotar, and Sydri-Fliotar. To these, however, which are totally uninhabitable, ought, perhaps, to be added the estate of Eystri-Dalbær, in Landbrot; for this farm, which has always been subject to annual damages, by the drifted sand from the districts to the eastward of it, is now, by reason of the disappearance of the waters of the Skaptaa, that formerly swallowed up a great portion of the sand, totally destroyed. Yet it has not received any damage immediately from the fire. Exclusive of the two cottages, called Kalfafellskot and Blomsturvellir, there are in all, twenty-nine farm-houses which the fire has more or less damaged, yet not to that degree, but they may in the course of time again be tenanted. Of these there are at present fifteen lying desolate, Hvammur, Svinadalur, Eystri-Asar, Botnar, Hnausar, Dalbær, Hollt, Skal, Selialand, Thvera, Nupar, Kalfafell, Mariubacki, Hvoll, and Skaptardalur. The great distance from the coast and the difficulty of approaching it, will, however, probably be insuperable objections to the last of these places again becoming inhabited. Of the houses that are damaged, exclusively of the fifteen farm-houses mentioned as being desolate, there are fourteen still inhabited, Buland, Ytre-Asar, Flaga, Hrifunes, Leidvollur, Langhollt, Stadenhollt, Efri-Steinsmyri, one of the five habitations in Sydri-Steinsmyri, Ytra-hraun in Landbrot, Eystri-Tunga, Kirkebai-cloister, Hunkurbackur, and Heidi. The number of farm-houses that are in part destroyed and partly damaged, amounts in all to thirty-seven, besides four cottages. Twenty-three only, out of this number, now remain uninhabited, which is rather more than one quarter of the eighty-nine farm-houses before noticed. It may farther be considered as a matter of consolation, that the damage sustained by the destruction of the grounds that are favorable to the growth of the Sea-Lyme-grass *(Elymus arenarius)* can scarcely be estimated at one half of the amount stated by Mr. Holm, and that, with a few exceptions, the whole of the sheep-walks in Sidumanna-afrettur (which in the large chart Mr. Holm has laid down as being entirely buried under everlasting lava, and has lamented accordingly in his statement) are yet safe, and may in fact be said to be untouched by the lava. I must, indeed, acknowledge that no small quantity of sand which is scattered over them, will render them barren for some few years; but it may likewise be truly said that they are become unproductive at a most convenient season, as the inhabitants are at the same time deprived of their sheep and cattle; for the loss would have been otherwise most severely felt. It will certainly be four or five, nay, probably, ten years before Siden, the Landbrotet, and Medallandet can again be so well inhabited as formerly; before the farms will again possess their full number of cattle; and before a sufficient stock of sheep can be reared. We have however strong reasons for supposing that, previously to that time, the pastures in general will have fully recovered themselves, and will be in better condition than they were at any former period.

§ XXI.

It has been already stated that, from the lava having filled up the channels of many of the larger rivers, as well as of the small brooks, these are in consequence totally stopped. Among the principal of them, we may reckon the Skaptaa, Hverfisfliot, Steinsmyrarfliot, Sydri-Ofæra, and Efri-Ofæra, to which might also be added the larger river, Tungu, or Kuda; for the current of lava had extended itself partly into its channel, and dammed up a portion of the river itself. The lesser streams that are stopped up, are those of Fedgaqvisl, Landa, Melqvisl, Hrossatuna, Hellisa, Laxa, Lingaqvisl, Fiardara, or Fiadra, Hollisa, and three small brooks in Landbrotet, Gloppulækur, Vordulækur, and Tungulækur, which however can only be considered as smaller branches of the Skaptaa, and necessarily all disappeared with the parent stream. But it may be asked, what has then become of all these rivers and brooks? To give a full and satisfactory reply to this question, and to ascertain the course now taken by the remnants of the streams, much local information is required, and it is above all necessary to be well acquainted with those parts of the rivers that were first stopped up, and with the obstructions arising from the natural situation of the adjacent country. I shall now endeavor to lay before the reader, all that it has been in my power to collect upon this subject, and shall first direct his attention to the eastward, beginning with Hverfisfliotet.

§ XXII.

Hverfis-fliotet. The upper part of the river Hverfisfliot, which still continues to run freely as far as the lava, becomes at this place a confined and stagnant water, the lava having not only entirely filled up the channel of this large river, but also extended itself to a considerable distance over its banks, so that the stream has been constrained to work its way under the superincumbent volcanic mass, visible only where it fills up such hollow places as it occasionally meets with. It is owing to this circumstance, that we find several pools of stagnant water below those places in the channel of the river that had been stopped: but no such are to be seen to the north of this place; and, as the vast quantity of water which was continually rushing down from the ice-mountains could not by any means make itself a sufficient outlet through the lava, the nearest hollows and crevices of which, had been already filled up with the impeded waters, it became necessary that some of it should discharge itself farther on, along the side of the rocky hills, and on the east of the lava; where a considerable broo is now, consequently, visible. This stream forced its passage in many places with great violence, especially to the eastward of the farm-house of Eystridalur, and thence continually ran in a direction parallel with that of the lava, stretching toward the sandy districts to the south. On the western side of the lava there was likewise a brook running past the farm of Thvera, and parallel with the lava. This, also, at length, empties itself into the southern sandy dis trict; yet it is by no means my opinion, that this stream can have taken its rise solely from the Hverfisfliot. Indeed, the color of its waters sufficiently proves a mixture of the mountain, and of the jökul streams; the latter discharging themselves into the plain from the jökul stream of the Hverfisfliot, which was choked up by the lava; whereas the other stream proceeded from various small rivulets and springs, which descended on the western sides of the mountains.

Without much reflection we might be inclined to believe that these waters would at length rejoin, and run along the original channel of the river; but it must be remembered, that the channel itself, was originally not deep at this place, but had been subject to many alterations and shiftings, and likewise that, wherever it stretched out into the sandy plain (being divided into different branches) it carried along with it, and heaped up continually, the loose sand. On the 22nd and 24th of July, 1784, when, in company with Mr. Livetzen and some other persons, I travelled over this district, we could scarcely observe any traces of the bed of the old river, and it is not a little remarkable that the whole of the vast quantity of water, which had here spread itself over the sandy plain, was still smoking in many places. In some parts so great was the heat, that we could scarcely bear to hold our hands in it; a circumstance that was rendered more particularly unpleasant by our being under the necessity of riding at a foot-pace for three long Danish miles through this hot, and, in general, very deep water. Every where, too, we were enveloped in a thick sulphureous fog and haze, that arose from the surface of it. In all probability the confined water will increase still more beneath the lava; particularly when it becomes thoroughly cooled, and nothing is lost by evaporation. A new channel for the Hverfisfliot will consequently be formed; for the water, now held in an unnatural state, from being as it were dammed up, must force itself a passage, either through the lava, or by breaking down its sides; yet I do not apprehend that any inundation will be the consequence, or that any damage will ensue, except indeed to the two farm-houses, Selialand on the eastern and Thvera on the western side. Till, however, such time as a new route is effected across the lava itself, between Siden and Fliotshverfet, and also between Thvera and Selialand, it will be very troublesome for persons who may have to travel between these two places; as they will be obliged to pursue a long and tedious course round the whole extent of the lava.

§ XXIII.

Skaptaa and other waters.

Having now arrived at the Skaptaa, and examined, not merely this great stream, but also the subordinate rivers and brooks, which had been stopped in their course or had wholly lost their waters, I found that this river still continued to flow uninterruptedly from its origin as far as a place below Uxatindur and opposite to Hordubreid, where the torrent of fire that had from the northward broken through Ulfarsdal filled up the channel, as has been before described. That the river has not been impeded in its progress before it reached this place, I conclude from the circumstance of its being here that we first observed stagnant water, which I consider a sure sign. It was not possible to have a view of the river itself, on account of the lava having, above this spot, accumulated to a considerable height; and because the atmosphere was every where filled with a thick smoke, caused by the dreadful heat, which still existed both here and throughout the whole extent of the upper part of the lava.

A little below this place the volcanic matter becomes still narrower, and there, on its western side, as well as upon the lava itself, we remarked several large pools of stagnant water, collected from the rivers Efri and Sydri-Ofæra, which, as well as the Skaptaa, were here choked up. Some streamlets, indeed, had found a passage along the west of the hraun, but nevertheless in this spot the confined waters became larger, and were more connected with each other.

Towards the east no body of water appeared to have made its way, until we came where the lava had filled up the channel of the Hrossatuna, compelling so much of that river as was not consumed by the fire-stream to work itself a new channel along that side. The quantity, however, was trifling, and no considerable current was to be seen, until the Hellisa, which was likewise stopped up, forcing itself a passage in a similar manner, united with the other at some distance above the farm of Skaptardal. On the outskirts of this farm the lava was remarkably narrow, and extended only a little way beyond the sides of the channel; but farther on it had spread over a great space, having its surface every where diversified with large bodies of water, many of which flowed one into another with a very strong current, and afterwards precipitated themselves in cascades, many of which were at once to be seen and heard. The Skaptaa broke forth with great violence on the western side. Even in the autumn of the year 1783, the lava on the outside of Skaptardal was so far cooled, that several persons who had fled from Siden ventured to pass over it to Skaptartungen, in order to avoid the tediously circuitous route from that place to Medallandet, and thence again across the Kudafliot. This passage they happily effected in safety, and they likewise conducted a number of sheep and cattle over with them. It is true that this took place late in the month of October, when the waters were frozen; but now, on the contrary, a large sheet of water was formed on the lava; all the hollows were filled up; and the progress of the water was impeded to such a degree, that only the highest summits of the hraun were visible. It will therefore be impossible for any person to cross over here so long as the waters are thus confined, unless indeed in the winter, in such times as they are a sheet of solid ice; and how far this will ever be the case I have great reason to doubt, there being so many strong and violent currents: nevertheless I will not wholly deny the possibility of its being effected. It may here not be amiss to state that the greatest quantity of water collected together was in the vicinity of Skaptardal; while smaller pools were to be seen in the more southern part of the lava; but in the eastern parts, as about Landbrot and Medalland, the quantity was inconsiderable. The whole of the water that fell on the north side at Skal was swallowed up by the lava: but this was no more than what proceeded from some small brooks, running down from the mountains; whilst on the other hand the Hollsta creek, which was stopped by the fire, formed itself into a large lake, which, being joined by the impeded streams of the brooks Fiadra and Laxa, forced a passage near the sides of the farms Heidi and Hunkurbackar. Hence it passed close to the houses, and at the last-mentioned place threatened the still remaining part of the adjacent field, as well as the houses, with total destruction.

Turning hence to the western side, we find an arm of the Skaptaa has made its way along the side of the lava-stream and surrounded Svartanup, a cottage belonging to the farm of Buland; after which, accompanied by the lava itself, it has inundated Littlanes, a cottage also belonging to the same estate. The farm-houses of Hvamm and Svinadal, very near which it ran, were much damaged by it, and between these it was in sundry places dammed up on the western side; thus giving birth in its course to various large lakes. Hence it ran forward with a moderately strong current, passing at no great distance the parsonage house at Eystri-Asar, the farm-house of Ytri-Asar, and the flying lands (as they are called from their being composed of drift sand) of Flaga and Hrijunes, where these waters, together with the Tungufliote, are stopped up, and, in the form of an exceedingly large inland lake, have inundated the neighboring country, doing infinite damage to the two last-mentioned farms, particularly Flaga, the meadows of which it has totally overflowed and destroyed. One arm or branch of this large lake ran, together with the Holmsa, into the Kuda-stream. In the Medallandet many dams were formed and many brooks dried up; among which may be mentioned Landa, Melqvisl, Fedgaqvisl, Lingaqvisl, and Steinsmyrarfliot. A quantity of impeded water, draining out from under the lava, has spread itself over the morasses belonging to the farm of Hnausar, and thus not only rendered these places more boggy than before, but likewise entirely surrounded the farm-house, so that for the present, at least, all access to it is nearly impracticable. The water, however, which has issued from the south-eastern point of the lava is by far the greatest impediment to travellers, and will probably in time become seriously dangerous, should it continue to increase in the same degree as it has done of late. We found it still smoking, though it could not be called hot: and observed where, having taken its course over deep morasses, it had in many places forced its way into the soil, forming holes in which horses might fall and be much injured, especially as the ground between these cavities is full of little hillocks and rugged places. All persons wishing to proceed from Alptaver or Medalland to Siden, must cross this stream, which already extends itself for a considerable length towards the sea with a moderate current. Its source is supposed to be one of the brooks that have been stopped up (probably Fedgaqvisl), which, after being confined by the lava for a time, has at length found a passage through it.

There is the greatest reason to apprehend lest the body of water here confined should hereafter swell to a size too large to be restrained within its present limits, and, uniting with the Skaptaa itself, should at some future time rush forward with destruction commensurate to their violence. The Landbrotet, I should hope, will always be secure against these inundations; but, on the other hand, the Medalland is exposed to the greatest hazard, since, immediately to the north of it, large quantities of water have been dammed up by the lava, and these at the time of my visit were evidently increasing, and the several pools were uniting one with another. When we were at the farm of Stadarholt, in Medalland, in the morning of the 28th of July, we heard a loud noise and splashing, which arose from the falling of water within the lava. Taking it for granted, therefore, that the Skaptaa itself, together with the other confined waters, must in the course of time force a passage through the lava, it follows that they must either overflow Medalland, to the great injury of that district, or must precipitate themselves into the Kuda creek, which will consequently be rendered impassable for people on horseback. The only apparent mode of crossing it therefore would be in boats or on rafts; for the union of the Skaptaa with the Kuda river, would not only make the body of water considerably increase in depth, but would also cause it to flow with a much stronger current than is the case at present, and thus, necessarily, render it impassable for horses. Even the expedient just mentioned might be attended with many difficulties; since the bed of the river consists of loose sand, which, by the force of the stream, is driven about, and formed here and there into large banks, over which it would not be found easy to pass with laden boats; especially when at the same time is taken into consideration the difficulty arising from the rapidity of the current. This, however, is stated merely on conjecture; and I am led into such remarks from the idea that these accumulated waters may force a passage at one or other of the two places just mentioned; there being no other obvious means by which they can make their escape. Nevertheless, as it is not given to man to penetrate into, or to anticipate, the hidden operations of the Deity, so we must hope and expect that the best result will happen in this, as well as in all other cases, from the superintendance and direction of his all-wise Providence.

§ XXIV.

It has been already remarked (§ XXI) that some branches of the Skaptaa, which formerly flowed through Landbrotet, as Gloppulækur, Vordulækur and Tungulækur, have, together with the parent fountain, been dried up. This district has not indeed sustained any great injury from the fire immediately; yet, nevertheless, since the disappearance of the Skaptaa, it is exposed to continual droughts, and may possibly in the course of time be totally destroyed by the flying sands from the eastern country. It is true it has always been subject to such disasters, but it had constantly in former cases great protection in the Skaptaa, which, by swallowing up the sand, prevented it from driving over to the western side, at least, in such a quantity as to effect any material damage. Now, on the contrary,

Flying or drifted sand.

most of the farms in this district are in the greatest danger; and so imminent is the peril to which Kirkebai-cloister is stated to have been exposed, even during the last year, from the drift-sand, that in case of its being annually revisited by similar misfortunes, it will in the course of a very short period become uninhabitable. When I travelled through this district, in the month of July last, the grass, in a great part of the enclosed pastures, was covered with sand, and large heaps of drifted sand lay between the houses, as well as scattered over the adjacent country.

§ XXV.

Some of the Phænomena attending the fire.

The principal phænomena attending this eruption have been already described. The thick smoke which by day issued from those districts that had been burnt, and at night appeared like a flame of fire, was still to be perceived in the month of March last, arising here and there from the lava. Since that period, however, the smoke alone has appeared. Whilst lying in my tent at Kudafliot, I noticed, not without wonder, the innumerable columns of smoke rising from the current of lava between Skaptartunga and Landbrot. These were particularly visible in three places towards the north, among the mountains; and I learned during my journey, that the most westerly arose from the channel of the Skaptaa, the eastern from the Hverfisfliot, and the middle ones from the source of the fire, and the district adjoining it. Beyond Skal our attention was excited by a very thick column, far surpassing all the others, which from this spot were to be seen rising by thousands, almost in a direct line with the burnt district towards the east, in forms innumerable and the most agreeable to the eye. Large bodies of smoke, together with some smaller columns, were issuing in various places from the eastern lava, near Fliotshverfet; but the smoke broke out in the greatest quantity between Nupar and Selialand, and also between Eystridal and Tholvardal, where the channel of the river had formerly been, and where, indeed, it might still be traced from the bay quite up to the place at which the river was first choked by the lava. Below this place the smoke appears to increase, contrary to what might be expected, that it would here be diminished by the confined waters of the Skaptaa, Ofærur, and other rivers and brooks: but the same circumstance was to be observed in the eastern branch, from the place where the Hverfisfliot is stopped up. The channel of the Skaptaa was particularly to be distinguished, as was that of the Hrossatunas, wherever the bed was filled by the lava. For a considerable distance above the farm of Skaptardal and beyond Næs, which was already burnt, there was in many places no smoke, and in other spots so little as scarcely to be perceptible; but below these places pillars of smoke were every where discernible along the whole of the southern current of the lava, but principally around Skalarstapi, where they were really dreadful. It is indeed asserted by several persons that this quantity has much increased since last spring, on which account many people entertained great fears lest a new eruption should take place. When in company with Mr. Levetzen and others, I performed a journey on the 26th of July, from Mörk to Siden, we turned off to go to Skalarfiall, a spot which afforded us a prospect equally pleasant and extraordinary; for we stood here upon the top of a very high mountain, which on the south side was entirely covered with grass. Here and there among the rocks some remains of the farm-house and church of Skal were still visible. From this place, which commands a very extensive view, we could see the whole mass of lava stretching over Siden and the Medallandet, and also a part of the western branch towards Skaptartunga. So extensive was the portion of lava towards the south, that the eye could not distinguish its boundary from that of the superincumbent clouds, in which its utmost extremity appeared to lose itself. As to the lava, it was every where of a coal-black hue. In its progress along the channel of the Skaptaa, as well as along several of its arms and auxiliary streams, it had formed itself into a number of lofty hills, running in a direction from east to west, and appearing from our elevated situation like a flight of steps. From each of these currents were rising in greater or less number, columns of smoke of different degrees of density, which appeared where we were standing to reach even to the clouds, exhibiting thousands of fantastic shapes. The black ground of the lava was to be seen between the lighter columns of smoke, which were, at a considerable height in the atmosphere, collected together into one thick bank of clouds, of a white or yellowish tinge, intermixed with shades of a deeper cast. To the south-west of this place was just discernible the upper part of the insulated and steep mountain of Skalarstapi, rearing its summit above all the surrounding lava, though at the same time almost enveloped in the dreadful smoke that was ascending on every part of it. The streams that issued from the rock poured down its sides, and added to the indescribable beauties of this enchanting scene, the effect of which was still farther increased by the view of the burnt farmhouse of Hollt. At Dalbae, in Landbrotet, which is likewise consumed, we remarked that one of the numerous eminences, shaped like a ball (which are so common in this neighborhood out of the reach of the fire, and which shew evident marks of having been in a state of conflagration at former periods), first began to smoke during the time of our residence in the eastern district. No smoke had ever before been seen to originate from these hillocks, nor was there any at the period of my first arrival; but it considerably increased while we remained in Siden.

§ XXVI.

Whether the fire was still burning.

From what has just been said respecting the numerous columns of smoke, it may perhaps be by many inferred that the fire was not yet extinguished; especially as the beds of lava, both in and beyond the valley, were, so late as the month of July, 1784, in a state of extreme heat. Nevertheless, when I travelled around the lava, I could not perceive any marks of fire, not even in the vicinity of the crater itself. Nor does it appear to me, that the dreadful heat still remaining in the lava, and the numerous columns of smoke arising from it, prove the contrary; for, with regard to the lava, nothing is more natural than that so vast a body, when once heated, should retain its heat long after it had ceased to be actually on fire, particularly when it has so great a depth as is the case in the channels of the rivers. Sir Isaac Newton has even ascertained that a red-hot metal ball, of only two inches diameter, will take about an hour to become thoroughly cool, and the same philosopher has laid it down as an axiom that the time required for cooling a body of this sort is in the same proportion as the squares of its diameter; therefore, although this rule may not be perfectly applicable to the burning lava, which is so loose and porous, yet some idea may be formed what an immense space of time will be requisite for cooling the new and dreadful streams of lava in the district of Skaptefield. The well-known Sicilian author, Massa, informs us, that in many places, in Catania, he has found the lava hot even eight years after the great eruption of Ætna, in 1669; and that the new mountain, which, during the eruption of 1766, arose in the middle district or temperate zone of Ætna, was still inwardly burning in the year 1770, four years after. Nay, farther, that the lava which flowed thence in 1766, retained a great degree of heat in the year 1770, in places where it was deep, and particularly where it had filled up ravines to the depth of two hundred * feet. Now, if the heat was so great in Sicily, at the expiration of four years, where the lava was not more than two hundred feet in thickness, can it be matter of surprise if the new branches of the volcanic torrent in Iceland, which in the channels of the rivers lay twice or thrice as deep, should still remain very hot at the expiration of only one year?

This may therefore be considered as a sufficient proof that the continuance of the heat in the lava is not to be regarded as a certain symptom of the continuance of fire. For my own part, I am thoroughly convinced that the whole was already extinct at the time of my travelling in the district of Skaptefield, which was in the month of July of last year; for, with regard to the number of pillars of smoke seen to arise, I consider them in reality as nothing more than vapors, produced by the vast quantity of water arising from the impeded rivulets,

* See *Brydone's 9th Letter.*

which every where, throughout the whole extent of the lava, become evaporated into steam. Such clouds of steam were more particularly abundant in rainy seasons, which makes it sufficiently clear that they owed their origin to water alone. When Mr. Levetzen and myself travelled from the farm-house of Skaptardal to the burnt farm of Skal, and thence up the mountains, on the 26th of July, we were enveloped in so thick a fog that we could hardly see for six yards before us. The fog had a disagreeable smell, and at length turned into a heavy fall of rain. But when we returned to Skal we were informed that this scent, as well as the fog itself, was to be found only on the hills, and that there had been, at the same time, fine and clear weather in the valley, together with a southerly wind, which had finally dispersed the rising vapors, carrying them towards the rocks in the north, and had at length enveloped the low grounds in a thick fog and heavy rain, similar to what we had previously experienced in the mountains. These circumstances tend still farther to confirm my conjecture that the smoke ascending from the lava is nothing more than water, converted into vapor by extreme heat, and consequently that it can by no means be regarded as a proof that fire still exists in an active state among the lava.

§ XXVII.

I now proceed to inquire into the state of the lava itself, and into the different kinds of this substance which have occurred to me. In doing this I shall first notice the color, which, whether at greater or less distances from the place of eruption, was every where of a greyish ash, intermixed with black: the latter more particularly predominating. In many places, where the lava presented an even surface, it had cooled in the same form in which it had flowed, and it was externally either deep red or violet, though the interior more frequently partook of a light-blue tint.

Of the nature of the lava.

Wherever the lava had, from the circumstance of various streams succeeding each other, formed itself into eminences, the heated vapors naturally forced their way through the upper and hardened crust, which was consequently broken, and had fallen down in fragments of various shapes and sizes. In places where the lava was smooth and even, it was from the like circumstance full of cracks and rents, and in such places, but principally in the deep hollows, it was strewed on the surface with a fine white kind of dust, somewhat resembling salt, of which, with infinite pains, I was able to collect a small quantity. In a few spots, also, where the surface of the lava was level, might be seen yellow veins of sulphur among the cracks. Some of this I endeavored to procure with a pick-axe, by which means I broke into a number of concealed vacancies, that exhibited a mixture of various colors; while from their roof or upper part hung a great quantity of small projecting points, either of a sulphureous yellow or a dark red color. At the extremity of these processes, were seen a quantity of red drops, which in drying had become indurated, but which, notwithstanding their small size, would bear a smart blow with a hammer without being broken. The bases of the cavities were chiefly yellow and green, though sometimes a reddish color was intermixed with them. In the bottom of one, I met with a beautiful kind of lava, remarkable for being most elegantly variegated with red, green, and yellow. Lava of the common sort, such as is every where to be found, has always a blackish hue within, with an exterior of a bluish cast, or sometimes with a mixture of red and violet color. Another species, somewhat different from the two just described, was found near Blæng, at a little distance from the lava-stream. A fourth kind of rock also is thrown out by the volcano, and carried across the Fliotshverfet. This appears to be a black slate, shining within like pit coal: its exterior is marked with numerous white dots. With regard to pumice-stone, I could find no large pieces of it. Those which I met with were about the size of a hazel-nut, of a brown color, fine in quality, its weight light, and its nature, as far as I could judge, exactly the same as that used in the arts. Under the five kinds of mineral substances here mentioned, which have been thrown up by the volcano, all others which have come under my inspection are to be classed, as they cannot be considered to differ essentially from them. For the information relative to their component parts, I am entirely indebted to the skill and knowledge of our celebrated countryman, Mr. Myhlensteth, who has had the goodness to make an accurate chemical analysis, not only of the various kinds of lava, but of the fine whitish substance resembling salt, which has been before mentioned, and also of various other substances, one among which deserves particularly to be noticed, a kind of ravelings, which resembles grey hairs. This was to be found every where in the vicinity of the fire, but especially on level sandy places, and no where in so great quantity as on the extensive sandy plain of Skeidarasand, to the eastward of Nupsvaters and Fliotshverfet. The filaments sometimes lay spread out singly upon the ground; sometimes mixed and interwoven with each other; in other places, twisted by the action of the wind into the form of garlands of various sizes, or into circles of a light grey color. The ravelings themselves, though short and broken, resembled the finest hair.

The following is the result of Mr. Myhlensteth's experiments upon the five different kinds of volcanic substances which were sent to him, and which have now been mentioned:

No. 1 resembles externally the scum of iron. Two ounces of it being pulverized, one ounce of iron was extracted from it, by means of the loadstone, and was perceptible, with a magnifying glass, in small thin laminæ, to which some lava was still attached. These two ounces were put into an equal quantity of oil of vitriol, mixed up with it, then thinned with eight parts of water. The vitriolic acid evidently dissolved the iron, when the solution was separated, which, at blood-heat, afforded a fine Prussian-blue color. The remaining part, that was not dissolved by the vitriolic acid, was thoroughly dried and found to have lost three quintins of its weight, and the loadstone had no effect upon it. Another experiment was tried with one ounce of this mineral, from which the magnet only extracted one and a half quintin, and the vitriolic acid dissolved only twenty-four grains of it.

No. 2. A mass of matter melted and run together, colored on the outside with red and green. It consists of sulphur, mixed with iron, copper, sand, and other kinds of earth which contain some acid of salt. If a piece of it is put into a moderate coal-fire, and thoroughly heated, it gives a very fine blue flame. Ten quintins of it were made red-hot in the fire, pulverized, mixed with an equal quantity of oil of vitriol, and diluted, during the operation, with eight parts of water. After some hours, the solution was separated, and gave a fine copper-color to a piece of polished iron. To this solution was added as much iron as could be dissolved, for the purpose of separating the copper, which, on being afterwards melted, was found to weigh fifteen grains. This copper yielded readily to the hammer, and with the addition of spirits of sal ammoniac afforded a very bright blue color. To the sediment, still undissolved, was again added half as much vitriolic acid, and the whole was treated in the same manner as before. The solution was then precipitated at blood heat, and afforded one quintin of Prussian blue. The still remaining sediment, when dried, weighed eight quintins, and consequently the acids had extracted one quintin and forty-five grains of iron, and fifteen grains of copper. The remaining eight quintins, when melted down with borax, gave a clear black glass, without regulus.

No. 3. A white kind of calcined earth, one ounce in weight. It caused an astringent sensation upon the tongue. On being put into boiling water, half a quintin of copper vitriol was extracted from it.

No. 4 was to all appearance nothing but common lava, and, as far as could be ascertained, consisted of sulphur ore mixed with iron, a little copper, and various kinds of earths, melted down together. The mineral acids had no effect upon the lava, and, on its being pulverized, a very small quantity of iron only was produced. Four small pieces, weighing three and a half quintins, which appeared to be more light and brittle than the rest, and totally free from white spots, were melted down with borax, when they yielded a lump of copper weighing three and a half grains. Half a pound of this lava was pulverized and properly melted, but it gave no more than three grains of copper.

No. 5 was a piece of common pumice-stone.

Of the whitish powder, with which the cavities of the lava were filled, I got a small sample weighing one and a half quintin. It had a saltish taste, and on crystallization afforded proper Glauber salt, which weighed two-thirds of a quintin, and ten grains of kitchen-salt.

The grey and hair-like ravelings abovementioned were found to be of the same nature as Nos. 1 and 4, and in all probability are the self-same substances drawn out into fine threads, which, from the delicacy of their structure, are easily broken, and are carried about by the wind in various directions to considerable distances. Twenty grains of them were melted down by means of a moderate fire to a black glass *. So

* Professor Wilke, at Stockholm, procured a small sample of this hair-like substance from Iceland, and has given a dissertation upon the subject, similar to

extremely brittle was the texture of the lava, even last autumn, when so far cooled as to suffer any one to hold it in his hand, that the application of moderate pressure instantly reduced it to a fine powder; but it was now on the contrary become considerably solid, insomuch that it was with great difficulty I could work through it, in many places, with the boring-instrument. Notwithstanding this, it was still very dangerous and unsafe, and was every where difficult to walk over. Unsafe it must also of necessity remain for a long time, on account of the numerous sharp points and projections with which its surface is covered, and upon which it is scarcely possible to tread with the thin Icelandic shoes, made of raw hides, or even with thick and properly soled ones, without immediately cutting them through. Much danger, too, arose from the circumstance of the hot vapors which it concealed, having, previously to the lava becoming thoroughly cool, produced, as above mentioned, innumerable hollow places, the arch-way or ceiling of which,

Solidity of the lava.

that inserted in *Dr. Crell's Annals of Chemistry, for* 1784. *Book* II. *p.* 323.

though in appearance resembling the rest of of the lava, yet was not sufficiently thick or strong to bear the human weight.

That the floors or bottoms of these cavities owe their colored appearance to the different kinds of sulphur, iron, and other metallic substances which have melted and dropped from the arched surface, will readily be perceived, without the necessity of my remarking it; and it will likewise easily be understood, that the spiculæ, which hang from the ceilings, are nothing more than a part of the lava, more or less intermixed with strong substances, which, whilst dropping, had cooled and become indurated.

§ XXVIII.

Attempts with the boring-instruments.

It has already been mentioned, (§ XVII.) that the great heat arising from the lava was no small obstruction to the experiments we had hoped to have made with our boring-instruments, in the hilly country, as well as in the vallies; insomuch, that I began to entertain fears lest this circumstance should render these instruments quite useless; preventing us, as it did, from employing them, except where the ground was proportionably cool. It is scarcely in the power of any one to form an idea of the difficulty that attended this part of our labors. To be continually turning the instruments round, and working through the hot, hard, and uneven lava, while we were at the same time treading upon its sharp-pointed edges, was certainly a task as painful as it was irksome. Nevertheless, by these trials, I found that the lava in some places did not lie more than six or eight feet deep; that in many it did not exceed ten feet; and that wheresoever, as was the case in certain situations, it was far deeper, its depth seemed wholly to depend on the peculiar nature of the country. Both below the lava, and close by its side, was found either sand or earth of the same kind as that which appears every where in this district, at a distance from the fire, as in the peat-bogs or in the grounds where the sea-lyme grass *(Elymus arenarius)* grows; but no kind of slate could be discovered in the neighborhood. The boring-instruments were useful, in enabling me to ascertain the quality of the soil below and about the lava, as well as the depth and nature of this latter beneath the surface of the ground, and they farther confirmed me in the opinion which I have stated above, that the volcano was to be considered as arising from a partial eruption, and not from any internal and universal ignition of the earth.

Height of the lava.

The height, to which the heaps of lava rise in the level country, is in some parts very considerable; particularly at Skalarfiall, where they have reached up to the rocks that project from the south side of the mountain: yet, nevertheless, were we to calculate its extreme height on the plains at an hundred feet (and even this is not quite a fourth part of what has been stated *), I am still persuaded we should greatly exceed the reality.

§ XXIX.

It has been already noticed in its proper place, that, after the first breaking out of

* See *Holme's Account of the Fire*, p. 19, where the height of the lava is estimated at seventy fathoms, or four hundred and twenty feet.

State of the weather after the 1st of June, 1783.

the fire, a great quantity of ashes, sand, and sulphureous dust was thrown over the adjacent districts, particularly those of Siden and Fliotshverfet. The long continuance of westerly winds, too, drove the sand-bank away from Skaptartungen to the places just named; and the vast quantity of burning sand falling around scorched up all the grass in the fields about Fliotshverfet to such a degree, that there were no means of support for the cattle, and the inhabitants fled from all the farms in this district, excepting the most easterly one, called Nupstad, which, together with the neighboring farm of Raudaberg, remained uninjured by the hot ashes. It is an undoubted certainty, that, if Providence should be pleased to grant better seasons to Iceland than the present, not only the parsonage of Kalfafell, with the cottage of Kalfafellskot, appertaining to it, but also the farm-houses of Nupar, Mariubacki, and Hvoll, will, in a very few years, be restored to their former condition; especially as the lava itself has not reached them. We may then reckon the number of farm-houses damaged at twenty-five, instead of twenty-nine (see

§ xx.), and make the whole amount, including those totally destroyed, to be thirty-three. It will easily be conceived, that, in proportion as the air became more and more filled with ashes, sand, sulphur-dust, and the smoke and aqueous vapors arising from the burnt districts, it must likewise become more fetid and unwholesome; and, indeed, even intolerable to those who were afflicted with the asthma, who, at such a time, could scarcely draw their breath. The presence too, in the atmosphere, of this mass of extraneous particles, obstructed in some measure the light and warmth of the sun, and caused to prevail, even at the height of summer, a most piercing and unnatural cold; in addition to which, what was still more unseasonable, a heavy fall of snow took place on the 11th and 21st of June. It was however soon melted. Almost all the new eruptions were accompanied by showers of hailstones, of an extraordinary size, equalling that of a sparrow's egg. These caused much damage and destruction to the grass, and nearly killed both men and cattle; but the mischief occasioned even by these was trifling, in comparison of what was caused by the heavy rains *, which, mixing with the sand, ashes, and sulphur, that had before fallen in immense quantity, incrusted the fields with a kind of black coat, somewhat similar to ink, but thicker, which poisoned the grass, and rendered the water stinking and unfit for use. Even the rain itself, in descending, became impregnated with sulphur and ashes, which sorely affected the eyes, caused a giddiness in the head, and was attended with pain as often as it fell on the naked body. The sun, from the impurity of the

* "During one of the heavy falls of rain," it is stated by Holm that, "there was observed, at Drontheim and at other places in Norway, and also at Ferroe, an uncommon fall of sharp and salt rain, which was so penetrating, that it totally destroyed the leaves of the trees, and every vegetable it fell upon, by scorching them up and causing them to wither. At Ferroe, there fell a considerable quantity of ashes, sand, pumice, and brimstone, which covered the whole surface of the ground, whenever the wind blew from Iceland; and the distance between these two places, is at least eighty (Danish) miles. Ships that were sailing between Copenhagen and Norway, were frequently covered with ashes and brimstone, which stuck to the sails, masts, and decks, besmearing them all over with a black and pitchy matter."

air, lost his splendor, and was shorn of his beams: indeed, it was very seldom that he was at all visible; and, when he was so, he appeared as a ball of glowing metal. The smoke covered the whole face of the island * for weeks and months together, so that seamen could not get sight of the coast

* This dismal atmosphere was not confined to Iceland; an obscurity in the air, and an unusual redness of the sun, were remarked also in England. In a copy of *Horrebow's History of Iceland*, now before me, is the following marginal note written by Mr. Sparrow of Worlingham Hall, a gentleman to whom I am happy in thus having the opportunity of acknowledging the obligations I feel myself under, for the ready access he has granted me to his invaluable library, and especially for the use of some scarce works relative to Icelandic History.—"An eruption of Hecla (as it was for a long time supposed to be) broke out again in the spring of the year 1783. In the month of May, of that year, I was in Holland, where the sun appeared for a great length of time to be enveloped and obscured in a thick dry mist; the cause of which was not then known. About the end of the year, two very large and luminous meteors astonished the world; they took a south-westerly direction, and were seen, apparently at the same elevation, and nearly at the same point of time, in the eastern parts of England, and the southern parts of Europe. They were remarked about seven or eight o'clock in the evening, within,

until they were close upon it; and in the hilly country the haze was so thick, that it almost entirely impeded the prospect. Such was the height to which, in the summer of 1783, the columns of smoke ascended, that they might be seen at the distance of thirty Danish (one hundred and twenty English) miles from the land, appearing like clouds in the air. The same thickness in the atmosphere continued until the middle of September in the same year; but, after that time, a prevalence of southerly winds happily brought with them a clearer air. It is remarkable, that in the summer of 1783, these winds had always been attended with the finest weather, contrary to what might have been expected, that northerly winds would have been required, to drive to sea and disperse the smoke arising from the southern side of the country; but at this time, although it is not to be denied that the southerly winds necessarily

I think, six weeks of each other, and about the middle of November. They approached so near to the earth, that I remember hearing a servant say, he stooped as one passed over him, fearful of being struck by it. They went with amazing velocity, and were soon out of sight."

impelled the smoke from the volcano into the interior of the country, yet they nevertheless were accompanied by a clear air and fine weather. The cause of so remarkable a phænomenon has been supposed to be a volcanic eruption arising from out of the sea, to the northward of Iceland, or, possibly, from the eastern bay of Greenland; since it has been observed, that the thickest darkness has uniformly been experienced, and the greatest quantity of ashes fallen, during the prevalence of northerly winds. How far this conjecture may or may not be well founded, I will not presume to say; for, although we see that some notice has been taken in the Berlin papers *, printed at Copenhagen, of a fire said to have arisen out of the sea, between Iceland and Greenland, yet that circumstance must for the present be reckoned among those which require farther confirmation. Nevertheless, I must acknowledge that it is not altogether destitute of probability; for, if the smoke which was spread over the country with northerly winds did not originate in a place

* For the year 1783, in No. 96, and others.

burning in that direction, it will not be easy to conceive whence it could proceed, unless it may be supposed that the columns before mentioned, as abundant in the district of Skaptefield, had, by southerly winds, been carried far away to the north, and were now driven back by the winds blowing from that quarter. When the winter of 1783 came on, the hazy weather was less perceptible, yet it was still observable for three days in November, and again once in December; on the 13th, the 29th, 30th, and 31st of January, 1784; then twice in February; as also in March and April, and in June, after which it prevailed almost daily in July, whilst I remained in the vicinity of the fire. Indeed, it could not well be otherwise, but that as soon as the thick vapors, arising from the districts filled by the lava, were dispersed by the winds, they must necessarily cause fogs and hazy weather in those places to which they were driven. After the prevalence of so thick an atmosphere as has been just described, it was remarked, at several periods during the following winter of 1784, that the surface of the snow was covered with very fine dust or ashes. Nevertheless, this mist or fog brought with it no putrid air, subsequently to the close of the summer of 1783, with the single exception of two days in the month of April, 1784, when a very strong sulphureous smell was perceived even at the Bishop's residence at Skalholt, and at many places in the district of Aarnes. During the time I remained at Siden, in the month of July last, the air had a strong smell of sulphur as often as the winds were northerly, and this was particularly intolerable about sun-rise; so that I could then with difficulty draw my breath, whilst lying in my tent.

I have already hinted that the damage which the pastures have sustained from the torrents of rain are incalculable; for these, accompanied with continual lightning and with the most dreadful peals of thunder, have not only in many places rendered the surface of the earth for ever unproductive, by dislodging from the neighboring eminences great fragments of rock, but have elsewhere materially injured vegetation by covering the ground with black dust, mixed with the hair-like ravelings, ashes, sand, and sulphur, thus poisoning the animals, and consequently cutting off the very life springs of the inhabitants. The feet of the poor cattle, and their heads, as far at least as their eyes, and the inside of their mouths, became changed, by their going into these pastures, especially such as are damp and marshy, to a sulphureous yellow color, and were filled with wounds and boils. The fire itself having much decreased since the beginning of the winter of 1783, the heavy squalls of rain became less frequent after that period, yet they nevertheless once or twice happened whilst I remained in the vicinity of the eruption. I shall never forget the last of these, which I witnessed while travelling with my companion in Skaptartungen. It was terrible in the extreme; and the rain fell, not in drops, but, as it were, in continued streams, accompanied with unceasing thunder and lightning; so that we were completely wet through in less than a quarter of an hour. We at length reached the parsonage-house of Eystri-Asar, which had been for more than a twelvemonth deserted, and there took shelter for an hour, till the tempest had subsided. Whilst here, we examined the church, but could not perceive that it had sustained any damage from the convulsions of the earth *. When we afterwards travelled across the heaths to the northward of this place, towards Svinedal, we noticed, with the greatest astonishment, that large lumps of ice, three inches in diameter, and one or two inches in thickness (which for the most part consisted of congealed hailstones), had fallen in various places during this dreadful rain. At but a short distance from the burnt districts, for instance, at Alptaveret, there was fine clear weather, so that these torrents of rain were only experienced in the immediate vicinity of the places just mentioned. I am therefore fully convinced that they were nothing more than the watery vapours which had arisen from the hot lava, and which (now that their weight overpowered the elasticity of the subjacent atmosphere) formed themselves into clouds and fell in torrents. It follows by the same mode of reasoning, that the rain will also tend to confirm the conjecture which I had long before expressed

* See *Holm's Account of the Eruption*, p. 21.

(§ XXVI.), that the smoke arising from the lava was only water, evaporated by excessive heat.

§ XXX.

Effects of the Volcano on the Fisheries.

The volcano likewise affected the fisheries in the year 1783; as the thick clouds of smoke and dust, which continually covered the land, rendered it too dangerous for the fishermen to put out to sea, and prevented their finding the proper fishing-stations. In consequence of this the summer fishery proved very inconsiderable. In the district of West Skaptefield the fire had a still greater and more destructive effect on the trout-fisheries, in the fresh-water lakes; for a larger quantity of volcanic ashes and sand had naturally fallen here than in other parts of the country, and these imparted an unusually blue color, sometimes tinged with yellow, to the waters, which at length became so foul and putrid, that great numbers of the fish were driven dead upon the beach. On the other hand, however, the drying up of the Skaptaa was of essential service to the inhabitants; because a number of large salmon, which had previously gone up the stream, were thus, in various places, prevented from returning. In such, therefore, of the pools as still retained their water, the inhabitants had an excellent supply, these serving as complete stock-ponds to the neighborhood. I purposely denominate this fish, salmon, although I am aware that in the district of Skaptefield it is called trout, which appellation is likewise given to it by Mr. Holm. I have never had an opportunity of seeing it myself, but, according to all the descriptions I have received of it, it could be nothing else than salmon: indeed, the weight alone of the fish affords a sufficient proof of what they are; the general average being from fifteen to thirty pounds, and the largest rising to at least ten pounds more. All the endeavors I could use to obtain farther information respecting this fish, by enquiries among the inhabitants, enabled me to collect nothing more than that it is the same kind as that which is usually caught in the white creeks, near Skalholt, and in the Borgefiord, as well as at Grimsaa, and in various other places. It has long been considered as a decided matter that no salmon are to be found in the eastern part of Iceland, and I am inclined to think that it is solely for this reason, that the fish caught in the district of West Skaptefield have hitherto passed under the name of trout.

Exclusively of the damage and loss occasioned to the fisheries by the fire, this calamity has likewise driven from the country various kinds of birds, that used to build their nests there; among which the principal are the swans. The inhabitants were well acquainted with the time that these birds cast their feathers, which was in the month of August, when the people used to climb the rocks and take a great number of them: but the sulphureous smoke and stench effectually banished them all; and the few eggs, that were found in the deserted nests, were so thoroughly impregnated by the smoke, as to be unfit for eating.

§ XXXI.

Influence of the fire upon vegetation.

What influence the volcanic eruption has had on the growth

of the grass, and the miserable consequences that have ensued from the failure of the latter, cannot be described in few words. It is easy to conceive that the progress of vegetation, in the district of Skaptefield, where the fields were immediately covered by the poisonous black substances, must unavoidably be stopped. But the misery was far from being confined to this place alone; for, even out of the district, where the volcanic sand and sulphureous ashes did not fall in any considerable quantities, the growth of the grass, which, until the eruption took place, was in a most promising state, was after this time totally prevented. Plants of all kinds withered, and became so brittle that the mere treading upon them reduced them to powder. The first that felt the baneful influence were the Buttercup *(Ranunculus acris)*, in Danish called Smörurt, and the Fisilen *(Leontodon Taraxacum)*. The Elting *(Equisetum fluviatile)* was the last to suffer. The same poisonous dust also attacked the cabbages and other vegetables in the gardens, totally checking their growth; and, having thus extended itself over the whole country, caused a general failure of the crops of grass. Not, however, equally in all places; for the want was particularly experienced in the northern district, where, according to report, the united produce of several farms at Langanaes was not more than sufficient to feed a single cow. It is true that the number of horned cattle and sheep was already greatly decreased, previously to the eruption, a circumstance which was partly occasioned by a succession of bad years, and partly by the infection that had recently prevailed among the sheep, and had induced a necessity of destroying great numbers. But still the loss was most severely felt; for, in the autumn of 1783, the natives were obliged to kill more than a third, nay, in some parts, even the half, of their remaining stock of cattle, for want of fodder. What is farther remarkable is, that in the summer of 1783, the pastures in many places swarmed with little winged insects, of a species hitherto unknown in Iceland. These were of blue, red, yellow, and brown colors, and appeared nearly to resemble the earth-fly. They were particularly troublesome to those employed in securing the hay, who

Insects among the grass.

were soon covered with their unwelcome guests. Many people have assured me that they even found numbers of them still living among the hay, in the depth of the ensuing severe winter; and, what is yet more extraordinary, that they left their quarters after a day or two of thaw or mild weather.

I have no reason to think that the thickness of the air had any apparent effect upon the grass in the late summer of 1784. The hazy atmosphere before described, which was occasioned by the smoke arising from the lava, was but seldom observed out of the district of Skaptefield, and the weather was likewise very mild and warm, from the latter part of the month of April till the end of July; yet still the growth of grass was almost every where indifferent, and the pastures occasionally frozen, especially where the soil was firm and level. Some hopes, however, are entertained, that, notwithstanding the very moderate crops of grass, a supply of fodder, however scanty, may be secured for the surviving cattle. The case is quite different in Siden and Medallandet, and perhaps in the whole of the district of Western Skaptefield; for there, provided the continued rains have not altogether prevented the hay from being harvested, there is no fear of a similar scarcity; the grass having grown in the greatest luxuriance, nay, even in an almost incredible quantity, both in the Medallandet and in Siden, and likewise on the two most easterly and deserted farms of Nupstad and Raudaberg, in Fliotshverfet.

I am strongly inclined to believe that this extraordinary degree of fertility is chiefly ascribable to the ashes, which have been thrown out by the volcano, and have fallen in the vallies, serving them both as the means of protection to the herbage and as manure. The great and rapid growth of the forests around Ætna * has always been attributed to a similar cause, and it has likewise been remarked in Iceland that a luxuriant vegetation generally succeeds the eruptions of Hecla †. This, therefore, in-

* See *Brydone's Letters through Sicily and Malta*, p. 89-93.

† See *Bishop Finsen's Account of Hecla*, 1766, p. 38.

duces the opinion that we must seek for the cause of the failure of the crops of grass all over the country, except in the places just mentioned, in the dreadfully severe frost * and cold of the preceding winter, when the earth was frozen to the depth of five or six feet; so that it was not entirely thawed in the beginning of the month of July, even in the neighborhood of the fire.

The loss sustained in this district by the destruction of the ground which used to produce the Sea Lyme-grass *(Elymus arenarius)* is the more deeply felt, since this plant has become an article of consequence among the inhabitants. The flour it yields is considered to be finer in quality and more nutritive than any which is imported †; so that, although the drying and preparing of

* In the winter of 1784, the thermometer upon Réaumur's scale varied from ten to twenty degrees of cold, and at Skalholt, Bishop Finsen once remarked Réaumur's thermometer at twenty-one degrees below the point of congelation. The excessive severity of that season continued till the end of the month of April.

† See *Olafsen and Povelsen's Travels in Iceland.* § 810.

the grain are but imperfectly understood in this district, it was nevertheless in so general use, that little or no other corn was bought at the trading towns. There are, however, notwithstanding the general calamity, some few of these grounds still remaining uninjured, and these, so early as the latter end of the month of July last year, appeared in a most flourishing state; for the remark, already made as to grass in general, holds good also with the *Elymus arenarius*, that volcanic ashes are its best manure.

In the district of Western Skaptefield, and especially at Siden, the *Hvannarot* (the root of *Angelica Archangelica*), the *Holl-tarot*, or *Hardasoe* (the root of *Silene acaulis*), and the *Gelldingarot* (the root of *Statice Armeria*), have also been used by the inhabitants as common articles of food, particularly in the spring, or in seasons of scarcity. They are also not unacquainted with the means of preserving their stock of Angelica root, which they gather in the autumn, and secure during the winter, by burying it a sufficient depth in the earth to be out of the reach of the frost, or by laying it in dry sand, from which they take a part as it is wanted for use, and eat it with butter. The other kinds of roots are generally dug up in the spring, and, as soon as freed from the soil, are eaten either raw, or boiled in water with a little milk. In the summer season a quantity of the *Lichen islandicus* (called in the Icelandic language, *Fiallagros)*, is likewise collected from the rocks for winter use: but immediately after the bursting out of the fire, in the year 1783, this plant, so important to the inhabitants, was, together with those before mentioned, which grew in great abundance in Sidumannna-afrett, buried under an immense covering of volcanic ashes, and coarse sand. Even to the present day the natives have to regret, in all parts of the country, that this Lichen, so valuable to the farmer, has not yet recovered itself.

§ XXXII.

Effect on animals. In consequence of the deficiency in the pastures, and particularly, of the poisoned state of the herbage, a great mortality naturally ensued among the cattle. In the district of West Skaptefield, where the

TABLE.

DISTRICTS.	*Died* 1783—1784.			*Remaining in* 1784.			*The former Number* 1783.		
	Horses.	Horned Cattle.	Sheep.	Horses.	Horned Cattle.	Sheep.	Horses.	Horned Cattle.	Sheep.
Northern Muhle	581	128	8718	383	364	4806	964	492	13524
Southern Muhle	479	88	7785						
Eastern Skaptefields *(a)*	210	27	1813						
Western Skaptefields *(b)*							415	226	3347
Rangervalle	4137	1120	32631						
Kiose	838	386	5000	223	499	1174	1061	885	6174
Borgefiord	1806	772	11581	432	759	1767	2238	1531	13348
Myhre	1453	795	7020	179	426	751	1632	1221	7771
Hnappedals	345	154	2966	120	174	507	465	328	3473
Snœfieldnes *(c)*	697	245	3616	339	388	3243	1036	633	6859
Dale *(d)*	973	370	9068						
Bardestrand *(e)*	405	147	3268						
Isefiord *(f)*	258								
Hunevand	2612	768	11149						
Skagefiord	3225	1127	16018						
Oefiord	1469	674	9314						
Total	19488	6801	129947	1908	3064	14400	8448	5917	59916

(a.) This statement merely belongs to two parishes in the district, Biarnanes and Hofells; no accounts having been as yet received from the other parishes.

(b.) What cattle have died, or what still remain, in the district of Western Skaptefield, we have not yet been able to ascertain: it is only at present known that the farm-houses burnt, damaged, or deserted, at Siden, as also at Fliotshverfet, Skaptartungen, and part of the Medalland, had, previous to the eruption, a stock of about the number stated, of which a very few are now remaining.

(c.) No information has yet been received from the parishes of Helgefields and Biarnerhavns, in this district, concerning the number dead or remaining.

(d.) Among the number of sheep stated to have died in Dale, the lambs are not reckoned. Of the number still remaining in this district, we can only speak to those in the two parishes, Staderhols and Hvams, which amounted to 93 horses, 151 cows, and 450 sheep.

(e.) From five parishes in the district of Barderstrand, Muhle, Flatoe, Ottardals, Sandlaugsdals, and Saurboes, no information has yet been received concerning what have died, or what are still remaining.

(f.) That 258 horses have died in seven vicarages in the district of Isefiord, is the only information yet received from that part.

fields were entirely covered with the infectious sand, ashes, and sulphur, mixed into a pasty consistency by the heavy rains; where the showers of red-hot stones and pumice had totally destroyed the face of vegetation; where a stinking and suffocating smoke, accompanied by tempests, continual lightnings, thunder, and noises in the air, heavy subterraneous reports and dreadful shocks of earthquakes, obscured the atmosphere; where a terrific stream of fire, a melted mass of lava, had urged its impetuous course; in short, where all the most fearful phænomena in nature had concentrated themselves, as it were, in one spot, it was common to see the animals running about the pastures as if in a state of madness; and I am credibly informed, that many of them, unable to find food, or even shelter to defend themselves from the surrounding horrors, in a fit of desperation, plunged into the fire. The cows were in many instances secured and fed in stalls, but the sheep and horses were dispersed in such a manner, that scarcely half of the original number could again be collected. All the quadrupeds of the island had thriven wonderfully, and gained strength, during the mild winter and beautiful spring of 1783, but this did not prevent them from dying off in considerable numbers, during the week or fortnight immediately subsequent to the eruption, with inflammatory diseases caused by the poisonous quality of the food. Such was particularly the case with the sheep, of which, in the district of Skaptefield, it was remarked that, whereas in Iceland they generally walk facing the wind, they now regularly turned away from it; naturally anxious to avoid the strong sulphureous smell, which the infected breezes brought along with them. As the cold, too, at a distance from the fire, was unusually piercing, they instinctively approached the current of lava, by which many of them were overwhelmed and destroyed, in spite of all the exertions that were made to save them. Nor was the situation of the cows and horses much better; for, although the disease was to them not equally fatal, yet they became excessively lean, and, even in the best season of the year, the cows gave scarcely any milk. It was the same beyond the West Skaptefield district, and, indeed, nearly throughout the

the whole island. It was still farther remarked in different parts of Iceland, during the summer of 1783, that the sheep, in direct opposition to the experience of the inhabitants, and to the supposed natural propensity of the animals themselves, avoided the dry elevated places, and even the heaths and commons, which most abounded in rich grass; and, as soon as they were driven up to the heights, snuffed at the earth and searched among the grass, but without tasting it: then immediately turning round, ran to the morasses and wet places. The cause of this I attribute to the circumstance of the ashes and sulphureous dust having had a more permanent influence upon the elevated pasturage, than upon the herbage in moist and low situations, where a proportion of the ashes and sand must have sunk into the water, and where, besides, the grass, when rain fell, must have been much purified and refreshed. It may possibly be objected to this, that the rain would naturally also produce the same beneficial effects in the higher grounds; but it is on the other hand to be remarked that the grass and herbage on heaths and commons, where sheep principally delight to go, is small and short. Consequently, as often as a heavy rain fell upon the ashes and sulphureous dust here collected, these were converted into a kind of paste which could not penetrate the soil; so that all vegetation was covered with it: whereas, in the morasses, this paste was gradually dissolved in the watery soil, and, as the grass in such situations generally rises to a considerable height, the mixture of ashes only affected the lower part of it. This I therefore consider to be the cause why the sheep, during the summer of 1783, uniformly sought the moist places; and it may farther be added, that they there in some degree found a shelter from the penetrating cold and frequent tempests, which are much more prevalent in the hilly country than down in the vallies.

In addition to the inflammatory disease just mentioned as so fatal to the sheep: so early as the commencement of autumn, 1783, when they were collected from the hills, several of them were found to be attacked with a distemper hitherto unknown to the natives. The poor animals could neither walk nor stand: their teeth were loose, so as to prevent them from chewing their food: their cheeks were full of swellings; and their joints were contracted. Towards Christmas the sickness began to shew itself in a still greater degree, even among the stall-fed sheep, and also among the horned cattle, which rendered it necessary for them to be slaughtered. Many, however, fell victims to the distemper much sooner than was expected, when the disease attacked them internally. Thus it was often found that the heart, liver, lungs, and kidnies of these miserable animals were covered on all sides with boils and ulcers: they were in some cases much swollen, in others quite destroyed and hollowed out: one of the kidnies was frequently considerably distended, while the other was proportionably shrivelled. The jaw-bones were perforated, as if they had been bored with an instrument, and the ribs were knit together in a most extraordinary manner. The bones were reduced to a substance resembling gristle, and even the hardest became at the joints so tender, that they might easily be separated from each other. When the entrails, that had been diseased, were boiled, they shrivelled very remarkably, and, if merely rubbed between the fingers, turned at once to powder. Of these particulars I was an eye-witness; for, when we arrived in Iceland, in the middle of the month of April, 1784, this plague was in its full vigor, and I can with truth assert, that the greater number of the cattle then alive on the island fell victims to the distemper during my stay there. Having said thus much concerning the sickness of the quadrupeds, I will only add, that it has been generally more destructive among the sheep than the horned cattle, and that there are some parishes, amongst which are Muhle and Rangervalle, and others in the west country, where the latter have been comparatively but little affected.

According to information that we have received, the disorder has in some degree made its appearance in the districts of Guldbringue and Kiose, and likewise in various places in the west country; but still its greatest ravages have been in Skaptefield, Aarnes, Borgefiorde, Myre, and Hnappedal, and, indeed, through the whole of the north of the island. From the east no intelligence has yet been received of its having broken out there. In some horses, which I had the opportunity of seeing during my journey to the place of the eruption, the distemper exhibited the same external appearances as in the other cattle; but the teeth in those that I examined were not yet become loose. It was a melancholy sight to see the miserable and deplorable state to which these poor creatures were reduced. In one instance, in particular, it was really astonishing how the wretched animal could walk, or even stand upon its legs, and yet its owners, in the confusion and distress, occasioned by their flight from the spot, were under the necessity of laying a burthen upon it. No striking external marks of the disorder were perceptible among the horses, out of the district of Skaptefield, but it has nevertheless prevailed there, if not as the sole cause, yet certainly in union with others, to produce a general destruction both among them and the horned cattle: many having died suddenly, when they had a plentiful supply of hay; others when in pastures where there was a sufficiency of grass, of which they were never deprived either by ice or snow. To our utter astonishment, we saw horses in the most miserable state of leanness, in the richest meadows, and even actually starved to death, having preferred eating substances the most injurious, such as the wood of houses, the hair from each other's coats, or whatever else was within their reach, rather than touch the grass of last year's crop, still remaining in the pastures. This appears to me to be a sufficient proof of the poisonous state of the herbage, during the year 1783; and, although the circumstance has not yet been investigated, I am fully convinced that the entrails of the horses, have been equally, with those of other animals, infected with the distemper. The few inhabitants, who had still left them some of the old hay, of the year 1782, preserved their cattle in a healthy and good condition; but even here, when the new hay came into use, the disease began to appear among them.

I have farther to remark, that, during the last summer, several of the younger beasts were recovered by feeding upon the new grass.

It might seem contradictory, were I here to assert, that the whole destruction among the cattle is to be considered merely as an effect of the volcanic eruption; because I have before stated, that, in certain districts, which were within the operation of the fire, no particular distemper has yet made its appearance. I must, nevertheless, still maintain my opinion, that the fire has mostly contributed towards it: since this was, beyond a doubt, the cause of the unwholesome air and frequent tempests, as well as of the failure of the crops of grass and hay, in the summer of 1783.

The cattle had, at the close of that season, become remarkably lean, and consequently, were rendered unfit to withstand the rigors of the ensuing winter, one of the most severe hitherto known. The inhabitants had not, by any means, a sufficiency of provender for them; nor were they aware, at first, of the unwholesome and poisonous quality of that which they did possess. It may be easily supposed, that the inclemency of the weather greatly contributed to the destruction, although the fire itself was the principal and original cause of it.

The extent of the injury which Iceland has suffered by the loss of her cattle it is impossible at present accurately to ascertain; as no correct statement has hitherto been made of what have died, or of what are still remaining. I annex, however, the following table, which is extracted from official information, and from lists that have been transmitted to the Royal Treasury, by the proper officers, merely for the purpose of specifying, though in a general way, a part of the destruction. This table, notwithstanding its imperfections, inasmuch as it does not extend to the whole country, and is besides too vague, and not sufficiently explicit in particulars for some of the districts, nevertheless, proves that, as Rangervalle, Skagefiord, and Borgefiord, had, in proportion to their size and population, the greatest quantity of cattle and sheep, of all the districts therein specified, so they have also sustained the greatest loss, and thence a similar inference may be drawn as to the parts unnoticed. According to the information that we have obtained, the northern districts have not suffered less than the rest, and their present deplorable condition may be put on a parallel with that of Western Skaptefield.

§ XXXIII.

Effects upon the human frame.

That the eruption had likewise a powerful effect on the human frame is certain, and is the less to be wondered at, as the unwholesome and pestilential air operating together with the noxious water and food, and with the want and distress occasioned by the destruction of the cattle, must naturally be productive of sickness and distempers. Diseases of the most inveterate kinds, in the form of scurvy, broke out in sundry places, and those even far distant from the fire: as, for instance, in the districts of Guldbringue, Borgefiord, and Myhre, especially in the first. The district of West Skaptefield was, however, the chief seat of this distemper; and in only six parishes there, no less than one hundred and fifty persons were carried off between the commencement of the new year and the month of June following; but some of these perished by famine. The same symptoms shewed themselves, in this disorder, in the human race, as among the cattle. The feet, thighs, hips, arms, throat, and head, were most dreadfully swelled, especially about the ankles, the knees, and the various joints, which last, as well as the ribs, were contracted. The sinews, too, were drawn up, with painful cramps, so that the wretched sufferers became crooked, and had an appearance the most pitiable. In addition to this, they were oppressed with pains across the breast and loins; their teeth became loose, and were covered with the swollen gums, which at length mortified, and fell off in large pieces of a black or sometimes dark blue color. Disgusting sores were formed in the palate and throat, and not uncommonly at the extremity of the disease, the tongue rotted entirely out of the mouth. This, dreadful, though, apparently, not very infectious, distemper, prevailed in almost every farm in the vicinity of the fire during the winter and spring; but, happily, its extreme horrors were confined to the district of West Skaptefield, beyond which it was attended with less melancholy consequences. Many of the unfortunate inhabitants, who resided in the vicinity of the place of eruption, and who could not procure either medicine or assistance, were starved to death; from an utter incapability of swallowing during the prevalence of the disorder any portion of food, even if they could obtain it, which was not often the case. On the farm of Nupstad, in the Fliotshverfet, which was the only one of all that remained inhabited, till the spring of 1784, the distemper attacked every individual among the inhabitants, not leaving a single person in health to assist and comfort the sick with the necessary attendance. Report goes even so far as to state, that several persons had been lying dead in their houses for a considerable time, before any intelligence of their decease could reach Siden, the nearest station; and that the information was at length conveyed by some travellers from the east country, who accidently stopped at Nupstad, and there heard from the few survivors of the distressing situation of the district. Both there, and at Horgsland, and, indeed, at some other places, it was necessary to burn the bodies upon the spot; since there were no horses left, and but few persons who were able to convey the deceased to the church. I ought indeed to add, that the circumstance of the earth being frozen to a considerable depth, as well during the winter as the spring of 1784, made a measure of this kind the more indispensible; the few that were free from disease being so enfeebled by hunger, that they had by no means strength sufficient to break up the indurated ground, and open graves for so great a number of bodies as now required interment. As often, therefore, as burial was at all resorted to, six, seven, eight, and even ten bodies were placed in one grave, and, for the sake of sparing exertions that they were little able to encounter, this was frequently so shallow as barely to allow a covering of earth above tbe lid of the coffin. That the air, from such a mode of interment, must soon become corrupted and dangerous for the human race, especially in the summer season, is a fact that speaks for itself.

It is necessary for me here to remark, that the disorder principally attacked those who had previously suffered from want and hunger, and who had protracted a miserable existence by eating the flesh of such animals (not even excepting horses) as had died of the same distemper *, and by having recourse to boiled skins and other most unwholesome and indigestible food. From respect to my readers I forbear to enumerate a variety of other things, which, as articles of food, were in an equal or greater degree nauseous and disgusting, and which, were I to detail them, would serve to show what shocking expedients the extreme cravings of appetite will drive men to have recourse to, and how that it is possible to convert almost every thing to food.

* I have been assured, in the district of Skaptefield, that the flesh and milk of sick animals had a remarkably unpleasant taste, and that, in particular, the milk was of an unusually dark and yellow color.

Some of the inhabitants, during the whole course of the winter, had not the least morsel of any kind of fresh or wholesome victuals, nor were they able to procure any other beverage than the water, which had been corrupted by the mixture of ashes and sulphur-dust. It was not all, however, even in this case, who died, but some recovered after having, in the course of the following summer, had a fresh supply of cows, and some provisions conveyed to them from the sea-coast, and after the pastures once more afforded them their wonted supply, being again covered with good grass and herbage, among which last were the various kinds of sorrel (*Rumex Acetosa* and other species) and the dandelion *(Leontodon Taraxacum)*, of which the natives made spoon-meat.

In my endeavors to ascertain the nature and origin of this distemper, I have not relied solely on my own judgment, but have solicited information on the subject from my valuable friend, our learned Professor, Kratzenstein, who deduces it from the same causes, and classes it with the same disorders, as Professor Callisen, to whose goodness I am indebted for the following remarks:

"The epidemic distemper, which broke out in Iceland in the vicinity of the volcanic eruption, appears to me, from all its attendant symptoms, to be entirely of a scorbutic and putrid nature, and exactly corresponding with the appearances which I have observed to accompany the highest degree of scurvy in cold climates. It undoubtedly owes its origin to bad provisions and water, and to the deprivations to which the unhappy inhabitants of the district were subjected. It is therefore most natural to suppose, and experience confirm the supposition, that no other remedy or relief could be found for these wretched people but a meliorated diet of fresh vegetables and fresh animal food."

§ XXXIV.

General consequences. The volcanic eruption having thus been productive of devastation and sickness, both among man and beast, a great famine and unexampled misery throughout the country, naturally ensued. The peasant, who, with the loss of his cattle, was likewise deprived of his sole means of subsistence, and of the best and most valuable part of his property, had nothing else (after having eaten the animals that died by famine and sickness) wherewith to satisfy the painful cravings of hunger, but skins and old hides, which he then boiled and devoured. Many, driven to the last extremity, have killed the few healthy cattle and sheep that still remained, and afterwards, when these were consumed, wandered with their whole families down to the sea-side, where they have become an intolerable burthen and source of impoverishment to the inhabitants of the coast. At the same time, too, that the uplands are become desolate, the condition of the inhabitant of the coasts is so much the more pitiable; as he can no longer continue his laborious toil through storms and frosts, with vigor and energy, unable as he is, to obtain the smallest quantity of butter or other strengthening articles of food to add to his present wretched fare; and being reduced to water, too, as his only drink; since whey, which was his usual beverage, is denied him. All this, as is known by long and sad experience in Iceland, renders the fishermen weak and dispirited, and unfits them for their ordinary occupations: thus, each hanging on each, the misery that began with one runs through all. The want of skins for sea-clothing will likewise for some years be a great obstacle to the carrying on of the fisheries with advantage; for although, since the mortality among the cattle, there is so great a quantity of hides in the country that they are considered as scarcely of any value, yet it is a well-known fact that those of animals which have died of hunger are in general unfit for use, and these, therefore, will neither answer the purpose of making coats or even of being manufactured into the shoes in use in the country.

The loss of the horned cattle and sheep was very severely felt by the Icelanders, but that of the horses was equally so, especially by the inhabitauts of the interior of the country, who thus found themselves deprived of their last resource, the means of having provisions and other necessaries conveyed from the coast, through long and tedious roads. Nay, many who are totally destitute of horses are under the necessity of carrying every load of hay into the outhouses upon their own backs, and frequently from a very considerable distance. Nor is there any prospect of these invaluable animals being soon replaced.

In the district of West Skaptefield, where a great proportion of the people had nothing, during the whole of the winter of 1784, but the most unwholesome food, and consequently became subject to the disorders which have just been described, numbers of people necessarily perished, and, out of seventy families that dwelt nearest to the fire and forsook their homes, not more than one half are still remaining in the district, the other thirty-five having fled to other districts, where a few of them have continued, while a part wandered about the country, and the rest are dead. It is now fully ascertained that the farms already burnt, damaged, or destroyed, were, at the commencement of the fire, inhabited by four hundred and nine persons in the whole.

With the exception of the district of Western Skaptefield, it does not appear that any part of the island has suffered so much as Tingoe, in the northern district, where a great mortality happened both among men and cattle, insomuch that (according to statements transmitted to the Royal Treasury) seven hundred persons lost their lives by famine and want. One hundred also have perished in Skagefiord, and three hundred and fifty-five in Oefiord. In the parish of Norder-Muhle more than one hundred died last year of the same disorder; and, if we calculate the number of those that have died in the district of West Skaptefield only at forty-five, the whole amount of those that have lost their lives by famine (of whom lists have been sent in) will be thirteen hundred. The general distress in the northern country has been exceedingly great, as it has also been in Borgefiord and Myrer, in the southern and western districts. It is, however, much to be dreaded that a still greater famine and mortality have visited the country, or at least particular parts of it, during the last winter, that of 1785, when it was scarcely in the power of man to alleviate the calamity *

§ XXXV.

The prevalence of violent earthquakes, as well as of the fire itself, and the extraordinary destruction occasioned by these in the district of West Skaptefield, are circumstances which are rendered sufficiently apparent in Fliotshverfet by the great chasms in the earth, which are there particularly abundant. It has been before remarked (§ IX.) that earthquakes were more violent before the fire broke out, but that from the period of the eruption they gradually subsided; so that in the year 1784 the shocks were weak and scarcely per-

* The total number of persons that have perished in Iceland, in consequence of the volcanic eruption, amounted, as the Etatsroed himself has assured me, to nine thousand. *H.*

ceptible, excepting only for two days before I left the country, the 14th and 16th of August. It was in the afternoon of the former day, between four and five o'clock, that the whole house at Inderholme, in the district of Borgefiord (where I was then staying) began to tremble; and, as we expected nothing else than that it would instantly fall in, we naturally ran out. When I looked up to the steep mountain, called Akrafiel, to the northward of the farm-house, I perceived its whole south side obscured by a vast body of smoke, arising from the fragments of rock which were continually falling. In another place, a little below the farm-house, large masses were broken off a lofty ridge of rock that rose near the sea; yet, thanks to God, there was no damage done at this place. On the following night several slight agitations were perceptible both to myself and other people then in the farm-house, sufficient indeed to rouse me from my sleep, though not to cause any serious alarm; but on the 16th we had again a long and dreadful shock, almost as heavy as the former had been.

This earthquake was most violently felt in the district of Aarnes, and has there also caused the greatest destruction, especially in the diocese of Skalholt, where, excepting the cathedral, only two small buildings are reported to have escaped without damage. All the rest, and among them the houses belonging to the college, were either entirely thrown down or materially injured. Some persons, who were buried in the ruins, were happily immediately dug out without having sustained any injury. The bishop, Mr. Finsen, and his lady, who, together with the rest of the inhabitants belonging to the episcopal residence, had long been obliged to lie under tents, in consequence of the constant succession of rain and tempestuous weather, were now reduced to the necessity of taking flight with their whole family; it being impossible to rebuild their palace before the winter came on. The timbers, in falling in, had broken, and were rendered unfit for use; nor were any horses to be procured for the fatiguing task of conveying fresh beams from the mercantile towns, situated at a distance.

The whole of the houses belonging to the Episcopal See of Skalholt having been in this manner destroyed, the University was necessarily neglected during the following winter. In the parish of Skeide, in the district of Aarnes, we are informed that all the farm-houses, two only excepted, had fallen to the ground, and that three persons in this district had lost their lives by the earthquakes. Besides which, these destructive earthquakes had every where caused great mischief, not only in this district, but also in the western part of Rangevalle, and, according to accounts that have been received, have damaged two hundred and fifty farm-houses on the estate belonging to the Bishopric of Skalholt, besides eleven churches, and have totally thrown down six other churches. On the other hand, both in the eastern district, and likewise at Vestmannoe *, as well as over the whole

* According to Mr. Sysselman Sivertsen's information, transmitted to the Royal Treasury, at the first shock which took place on the 14th of August, large rocks were torn from the mountains, and fell down on Vestmannoe, which was covered with smoke from the base to the summit: and, as the smoke arose from

of the south country, although they have been very perceptible, yet they have not caused any great devastation. How far they may at the same time have been felt with any violence in the district of Skaptefield is not yet known here: thus much only we can say with certainty, that some slight shocks had been perceived in the beginning of the month of August, at which time, the smoke appeared to have gathered strength in the wild and mountainous districts to the northward of Siden.

I am well aware that many people may be led to conjecture that these earthquakes must have proceeded from great revolutions in the bowels of the earth, or even possibly from the circumstance of new eruptions having taken place in the vicinity of the for-

several places at once, it was natural for the neighboring peasantry to be in great apprehension of more general destruction; but, nevertheless, no other remarkable damage appears to have been sustained than that of the largest and most valuable part of the *Bird-mountain* (a hill of the greatest value to the inhabitants), having been cleft and thrown down, and consequently rendered unserviceable for lodging the nests of the sea-fowl in future.

mer fire, and therefore must in that district have caused the greatest destruction. But, for my own part, I should rather be tempted to believe that, as these latter shocks were most violent in the district of Aarnes, weaker in Rangevalle and other southern districts, and so slight as to be scarcely perceptible either in the northern or the western parts of the island, that they owed their origin to some internal commotion in the earth, in the vicinity of Hecla; if they are not (which God forbid) a prelude to an eruption of the mountain itself.

It has also been shewn, that the annals of Iceland cannot produce an instance of an earthquake equally destructive as that just mentioned, which, exclusively of its having in a manner destroyed whole parishes and districts, has also reduced many of the inhabitants of the district of Aarnes to the most deplorable state, as the small stock of meat, and particularly of the common articles of food, such as butter, &c., which they had with the greatest difficulty secured during the preceding summer, were by this deplorable calamity spoiled by being buried under the ruins of the habitations: but above all it is to be lamented that this misfortune should take place just at the season of hay-making. The want of horses, too, is a circumstance very distressing to the country in general, and to the places destroyed by the earthquakes in particular, since, as observed above, without the assistance of these animals, the inhabitants can neither procure the timber necessary for building, nor any supply of provisions from the sea coast. It is therefore much to be feared that several of the farm-houses that are damaged must, for the present, remain uninhabited; especially as the hay has been almost entirely destroyed by this sudden misfortune, and by the long continuance of rainy weather following almost immediately upon it.

A consequence of these severe earthquakes has been, that the face of the country appears to be heaved up in the form of billows, and during the continuance of the shock it looked as if covered with a dark cloud of dust. All waters, as well the flowing as the stagnant ones, were sensibly disturbed and became white as milk; but the rivers themselves resembled the most furious millstreams. Many Hverar, or boiling-springs, and other brooks and pools were dried up, though some of these after a while again made their appearance in fresh places. The hot-springs about the Geyser, and above all the Geyser itself spouted out its torrents with a fury never before witnessed, and the same was also the case with the springs of this kind about Skalholt. It is very remarkable that, in the very place where I bored into the ground between these spots last year, there has sprung up, according to Bishop Finsen's account, dated the 14th of August following, a fresh fountain of boiling water.

We are also informed that the pastures in the district of Aarnes had, by these shocks in the ground, suffered such convulsions, that all the moss growing in damp places was forced out of the soil, and lay so thick upon the grass that scarcely any more hay could be cut; whilst in hard and dry places great cracks and apertures, nay, in some spots, even deep holes were formed in the earth.

§ XXXVI.

A new island arising from the sea.

In conclusion I have to add, that for a whole month previous to the volcanic eruption in the district of West Skaptefield, in 1783, a great fire was seen arising from the sea off the south-west coast of Iceland, and this was rendered visible to mariners, at the distance of six or eight Danish miles, by the vast body of smoke that proceeded from it. The sea around for twenty or thirty Danish miles was filled with pumice-stones to such a degree that they were no small obstruction to the progress of shipping. Of these pumice-stones, which were driven upon the southern coast in great quantity and in different places, I myself have found several here and there at Akranes, in the district of Borgefiord, and principally at Inderholme. But this is not all; for, by the force of the subterraneous fire, a new island has arisen from the sea, which was seen throwing out a vast quantity of fire by some mariners on their passing this coast early in the month of May, 1783. By the nearest estimate they could make, this island lay in about 63° 20″ of north latitude, and in about 354° 20″ of longitude, at the distance of seven or eight Danish miles south-west by the true compass from the outermost of the Fugle-skiers off Reikanes. Masters of vessels, who have sailed very close to this island, do not agree in their reports concerning its extent, some of them having calculated it at one mile in circumference, whereas others have described it as being only one-third of a mile or very little more. The island * is stated to consist of high rocks, in the rifts of which in two or three different places was burning a strong fire, which at intervals, as it burst forth, threw up a considerable quantity of pumice-stones.

At about one and one-third Danish miles by the compass from this place a sunken rock † was also discovered, over which the

* By later accounts we learn that this island was in the course of a twelvemonth reduced to a sunken rock, extremely dangerous to navigators. It is mentioned at p. 8 of this Journal. *H.*

† As I have not in any other work met with information respecting this sunken rock, it seems to me not

sea broke very heavily. By soundings taken when near the island it was ascertained that, at the depth of forty-two fathoms, the ground consisted of a kind of calcined stone-dust, which shone like pit-coal. At one place they had more than one hundred fathom of water, when only at the distance of half-a-mile E. N. E. from the island.

This island, to which His Royal Majesty has been graciously pleased to bestow the name of Nyoe (New Island) has not been seen this year by mariners: and though the ships in which Mr. Levetzen, Mr. Bulow, and myself went to the country and returned to Copenhagen, had express orders to search for it, we were still unable to discover it; notwithstanding that during our outward bound passage we continued cruising backwards and forwards for a long time in the latitude where we might expect to fall in with it. So that, if I may be permitted to draw any conclusion from this circumstance, it would be this: that Nyoe has sunk

unlikely that it is only the remains of the island just before described, which, as will hereafter be mentioned, is now scarcely to be seen at high water. *H.*

into the sea in the same manner as it rose a year ago."

While engaged in preparing this part of my work for the press, Sir Joseph Banks has been kind enough to send me a valuable Danish publication on the coasts and harbors of Iceland. It was printed at Copenhagen, 1788, and is entitled *Beskrivelse over den Islandske Kyst og alle Havne fra Fugle-skiœrene og til Stikkelsholm i Bredebug-ten med Forklaring over deres Indseiling*, ved P. de Löwenörn. From this I shall extract not only that part which concerns the New Island, mentioned by Mr. Stephensen in the beginning of the last section of his pamphlet, but also that which relates to the whole of the Fugle-Skiærene, as I consider the account of them too interesting, and the nautical information relative to them too important, to allow either of these to be omitted.

"From Cape Reykenes five single rocks, rising above the water, stretch out to the S.W. by W., by the true compass, and are called Fugle-Skiærene (or the Bird-rocks). The one which is nearest the land, and lies close under Reykanes, is called Carls-klippe: it is very dark, and has much the appearance of a church with pointed steeples. The distance between this and the second rock, called Eld-Ey * (or the Flour-bag), is one and a half Danish miles. Between these islands is the best channel and that which is most generally used. One may likewise pass between the other Fugle-Skiærs, if there is a tolerably fresh breeze; but the sea breaks very heavily, especially in spring tides, and may cause broken seas and put the vessel to great danger.

* *Eld-Ey*, or *Ild Oe*. The Icelanders call these rocks by the general appellation of *El-Eyranne, or Ild-öerne* (Fire islands), probably thereby intending to intimate that they have formerly been volcanoes, and have been produced by revolutions similar to those that have happened in the East Indies, in the Archipelago, at Sicily, and many other places, and very lately in Iceland 1783, with the *Blinde Fugle-Skiær*, as it is called; which, although it afterwards sank again and therefore justly bears the name of the *Blinde Skiær* (that is, sunken rock), may probably by some future convulsion again raise itself high above the water. More will presently be said concerning this *Blinde Fugle-Skiær*.

If opportunity offers I should always consider it safest to go between Carls-klippe and the Flour-bag, whether in coming from the eastward to the western harbors in Iceland, or in going from Iceland to the southward; both because the course is shorter and there are more certain sea marks. When clear of the Fugle-Skiærs, you must be on your guard, more especially if you turn to windward, against a dangerous sunken rock, called the Blinde Fugle-Skiær, of which I shall immediately have occasion to make mention.

I have laid down the Fugle-Skiærene, with regard to their situation between themselves and from Reikanes, according to Minor's description, with a few inconsiderable corrections from M. de Verdun's observations, and from a great number of bearings which I had the opportunity of taking, both when going to Iceland and on my return.

The outermost of the Fugle-Skiærs, which is called in the Icelandic language, Gier Fugla-Skiærdrange, and by Minor, Grenadeer-Huen (the Grenadier's-cap), lies, as nearly as can be ascertained by bearings taken from the sea, five and three-quarters Danish miles s. w. by w. by the true compass from the point of Reikanes, and consequently in 63° 44′ 40″ latitude and in the longitude of 25° 35′ 40″.

Lieutenant Grove has, near this place, had an observation in the latitude of 63° 44′ 20″ and on my homeward voyage, in sailing past it, I likewise had an observation of latitude and longitude, which answered very correctly to it. It is true that by my observation it lay a couple of minutes more to the southward in latitude, and the difference in longitude was likewise a couple of minutes, as it appeared to have been laid down too far to the eastward: but I have nevertheless left it unaltered with regard to the distance it is found to be from Reikanes, which must otherwise be corrected accordingly. It cannot be expected that observations taken at sea should correspond to so great a nicety, especially as the weather was not very favorable; but nevertheless I would not omit making this remark.

During the time that I remained at Holmens-Havn, Lieutenant Grove went out with a vessel under his command, for the purpose of navigating about that spot where the volcano island had made its appearance, in the year 1783, in order to discover if it still existed, or if any vestiges of it remained: but he found nothing but that which is called the Blinde Fugle-Skiær.

According to several very probable and well-founded suppositions, we have concluded that this is precisely the same rock which, in the year before mentioned, threw out fire, and cast up so much pumice-stone, that the navigators who passed the place found the sea covered with it. So long as it continued burning, it appeared above the water like a small island, which, as we learn from the statements given by mariners, who saw it that year, frequently altered its appearance; a circumstance undoubtedly occasioned by the lava and pumice-stone issuing from it; though it is probable that these substances have not been able to fix themselves firmly, but were washed away again by the sea breaking heavily against them; so that by these means the island that had started up, disappeared and was not to be found the next year, when orders were given to the outward bound ships to look out for it. The existence of the Blinde Fugle-Skiær, indeed, has been for some time known, but its situation has been so uncertain, that many people have gone so far as to doubt whether it actually existed, because they might often sail past, and even cruise about, without happening to see it. It is nevertheless extremely dangerous; and it is a most important matter to ascertain correctly where it lies, in order that we may be enabled to use the needful precaution in avoiding it. At the flow of the tide it is not visible, unless there is a sea running sufficiently high to break over it, and even then it is necessary to be very near to perceive it; but in the dark or in hazy weather it would probably not be possible to avoid it, should one be so unfortunate as to fall in with it. At low water, and when the sea is running off, about a cable's length off it may be seen dry. The sea breaks for the length of two cables. Round about it, the depth of water increases rapidly, and at the distance of from two to eight cables' length from it, the lead has shewn from twenty-six to forty fathoms, with small burnt stones resembling lava.

Lieutenant Grove observed the course and distance from thence to the Grenadier's Cap, or the outermost Fugle-Skiær; and when, on my return homeward, an opportunity offered for me to sail through the channel, I took numerous bearings to the Grenadier's Cap, and thereby ascertained my distance from it as correctly as it can be done at sea. I then shaped my course directly for the Blinde Fugle-Skiær; kept the log going; steered with the utmost diligence; and found the course from the outermost Fugle-Skiær to it, to be exactly the same as is laid down by Lieutenant Grove, 47° from the south to the west by the true compass, and the distance just four Danish miles; consequently, according to the situation of the Grenadier's Cap, it lies in 63° 32′ 45″ and 26° 2′ 50″. With clear weather, and especially if on board a tolerably lofty vessel, when between the two, this rock may be seen, or the breakers upon it, just at the time one gets sight of the outermost Fugle-Skiær; but if the weather is in the least degree hazy, the vessel would be too far from the Fugle-Skiær to enable a person to see it, so long as the Blinde Fugle-Skiær was in sight. When I approached the Blinde Fugle-Skiær I determined, according to the directions Lieutenant Grove had given me, to steer directly for it, and, although we consequently were continually in expectation of seeing it, yet we did not discover it until we were only at the distance of a few cables' lengths, when we saw the sea breaking over it.

Notwithstanding that I had not an observation for the variation of the compass, when close to the Fugle-Skiærene, yet I can judge nearly to a certainty from other observations, that, in the year 1786, it was from 36° to 37° north-westerly: and, as in the same year, I found it immediately on the western side of Shetland to be 26°, it consequently follows that the variation between Shetland and Iceland is, as nearly as can be calculated, $\frac{1}{2}^{\circ}$ for every degree of longitude we go to the westward. The variation increases very much afterwards to the westward of Iceland, and likewise when steering to the northward. I have observed the variation in Faxe Bay, and found it to be in the interior part of it from 37° to 38°, and, at the outer extremity of the same bay, from 38° to 39°; still higher, off Staal-bierg, the northern point of Brede-Bughten (Broad Bay), it was 40° direct westerly. In the channel, under 65° of latitude, and 35° of longitude, I found the variation by a series of observations, to be 45° 10′."

END OF APPENDIX. C.

APPENDIX. D.

ODES AND LETTERS

PRESENTED

BY THE LITERATI OF ICELAND

TO THE

RIGHT HONORABLE

SIR JOSEPH BANKS

AND THE

HONORABLE CAPTAIN JONES.

APPENDIX. D

ODES AND LETTERS.

THE ORIGINALS OF THE ODES WERE WRITTEN BOTH IN LATIN AND ICELANDIC; THE LATTER, HOWEVER, I HAVE NOT THOUGHT IT NECESSARY TO INSERT, EXCEPT IN THE INSTANCE OF THAT TO CAPTAIN JONES, WHICH IS GIVEN AS A SPECIMEN OF THE LANGUAGE OF THE COUNTRY.

Nos. 1, 2, and 3 are Addresses to SIR JOSEPH BANKS from BIARNE JONESON, and No. 4 a Letter from THEODORE JONESON.

No. 5 is a congratulatory Address from MAGNUS FINNUSEN to CAPTAIN JONES.

N° 1.

HEKLÆ VALE ANGLIS HEROIBUS.

I.

Auscultate,
Excelsi Jökli *,
Montes, tesqua,
In terrâ glaciali!

Me
Sapientes viri
Ambierunt quinque,
Omnes Britanni;

* Montana glacies.

II.

Inclytus Banks,
Inclytus Solander,
Cum pulchro
Comitatu.
Nunquam antea cumulata sum
Prioribus seculis
Tanto honore
Ab Anglicâ gente.

III.

Juvenis priùs eram,
Nemo favebat
Elegans vir
Annulatæ Virgini;
Sola steti
Longo tempore,
Donec Vulcanus
Me exornavit *.

IV.

Hìnc Nomen meum
Jam per terrarum orbem
Valdè inclaruit
Apud honestas nationes.
Multi desiderant
Antiquam grandævam
Oculis usurpare
Et ulnis complecti.

V.

Me cùm viri elegantes
Convenerunt,
Amorem exhibui
Juvenibus Dominis.
Perreptavit me
Flamma amoris.
Nullum denegavi
Viris honorem.

VI.

Monstravi illis
Rubra Cimelia,
Pluresque alias,
Quas habui,
Eximias opes,
Quas olim nacta sum

* Primam Heklæ eruptionem quidam Annales ponunt ad annum 1004; alii alitèr; si autem quot vicibus talia in monte hoc contigerint incendia scire desideras, vicies ter id factum esse creditur. Ab hoc tempore nobilitari imprimis cæpit, antea minùs celebris.

Ex Vulcani
Admirandâ fabricâ.

VII.

Me sapientes
Manibus contrectârunt
Antiquam Virginem,
Dederunt et oscula.
Bene sit
Alacribus viris,
Qui me inviserunt
Ex australibus oris!

VIII.

Resideo jam
Tristi fronte
Vidua desolata.
Sæpe Lachrymas fundo,
Postquam mei
Insignes amici
Reliquerunt me
Fortunâ orbatam.

IX.

Longum dolorem
Corde premo.
Neminem habeo
Cui aperiam.
Si vero vocem sopitam
Altiùs extulerim,
Res mira videbitur
Et immanis strepitus*.

X.

Vale Banks!
Vale Solander!
Valete omnes
Alacres viri!
Nolite oblivisci
Annosæ virginis
Reduces
In Angliam.

XI.

Largior ventos secundos;
Largior fortunam;
Largior nomen
(celebre);
Largior splendorem.
Sedete hilares
Ad compotationes†!

* Alluditur hìc ad horrendos et altisonos Heklæ strepitus in eruptionibus.

† *Guma Minni (verba archetypi Islandici)* propriè denotat memoriales scyphos, clarorum virorum in compotationibus

Bibite Nomen
Virginis, Eloquentes!

XII.

Nunc ad finem deducta
est
Hæc cantiuncula;
HEKLÆ VALE
Carminis titulus sit.
Date Versificatori
In præmium poematis
Dotem
Mihi convenientem.

evacuari solitos: hìc autem ipsa symposia.—Sub Ethnicismo certa pocula certis Diis, Regibus, vel Heroibus, consecrabant Veteres, qui mos etiam post introductam religionem Christianam, præsertìmque inter solennitates nuptiarum, apud nos religiosè fuit observatus. Sed, eliminatâ superstitione Papisticâ, in desuetudinem jam abierunt Scyphi isti memoriales; ut *Gudsrodurs minni*, Dei Patris Scyphus; *Heilags anda minni*, Sancti Spiritus Scyphus; *Mariu minni*, Beatæ Virginis Scyphus; *Marteins minni*, Martini Turonensis Scyphus, et id genus alia.

N° 2.

PRO FELICI IN ISLANDIAM ITINERE ET IN PATRIAM REDITU

MAGNATUM BRITANNORUM,

ANNO MDCCLXXII.

VOTUM.

Ludat hyperboreo Titan luculentus Olympo,
   Lunaque nocturnos clara gubernet equos,
Et vehemens Boreas pluviæ frigusque facessant!
   Cunctaque disfugiant, quæ nocuisse queunt!
Omine dum fausto Magni celebresque Britanni
   Observant Thules vasta theatra soli;
Quæ contemplari felix mens enthea gestit
   Perspiciat: clarum Jova secundet iter.
Quo bene confecto, tandem felicitèr omnes
   Restituat patriæ cura paterna Dei!

N° 3.

VIR CELEBERRIME!

Quòd me cum erudito tuo comitatu invisisti, grates ago quàm maximas. Mitto tibi jam, Vir humanissime, Carmen gratulatorium, paulo correctius auctiusque quàm antea. Nimis quidem exiguum hoc est munusculum, longèque tuam infra dignitatem positum, sed velis nihilo minus benignâ id suscipere fronte, inque meliorem partem interpretari, animum potius datoris quam doni vilitatem respiciens ; quâ de re eo certiorem spem foveo, quo evidentiora humanitatis tuæ habeo indicia ; velim id typis vulgari permittas, in Angliam, Deo duce, cum redieris. Adjici etiam posset Versio Anglica, si ita visum fuerit ; ut vestra in gentem nostram merita eo clariora evadant, atque hoc pacto in vulgus emanent. Iter vestrum ad Heklam quomodo cesserit scire gestio ; utinam bene et ex animi sententiâ.

Deus te salvum et sospitem patriæ reddat, omnesque tuos gloriosos conatus secundet.

Vale, Vir humanissime, nostrique memor, Felix diu vive.

Tui Nominis observantissimus,

BIARNUS JOHNEUS.

Schalholti, IV. *Calend. Octobr.*
Anni MDCCLXXII.

TRIPUDIUM,

A Musis Schalholtensibus agitatum
In Adventu
Celeberrimi Herois
DNI. JOSEPHI BANKS,
Armigeri,
Qui ex Angliâ in Islandiam transfretavit
ad eruenda et observanda quæcunque intra
Pauperculæ hujus Insulæ limites in Regno
Naturæ observatu digna,
Unà cum ornatissimo et eruditissimo comitatu,
Cum Historiæ naturalis Doctore,
Magni Nominis et solidæ Eruditionis viro,
DRE DANIELE SOLANDER,
Cum Astronomo, cum Antiquario,
Cum tribus Pictoribus, duobus Scribis,
Capitaneo navali, et subcenturione,
Honoris et debitæ observantiæ ergo,
Anglo Heroi ejusque Comitibus oblatum
Schalholti, A^o MDCCLXXII.
x. Calend. Octobris.

A BIARNO JOHNEO,
Philosoph. Mag. et Schol. Schalholt. Rectore.

I.

Fausto omine adsis
Cordata gens ab oris Angliæ.
Primùm tibi gratulabunda assurgit
Glacialis terra cum applausu !
Mariti, uxores, liberi,
Gaudio perfusi tripudiant.
Cælum, Salum, Solum, Solitudo,
Resonant cum lætitiâ.

II.

Præstantes olìm (Islandi),
Relictis patriis oris, Londinum studiosè petebant,
Artium addiscendarum cupidi,
Quas contenta libris cruditio commendat.
Oxoniæ in Anglico solo
Pedem hoc temporis tractu figere
Imprimis arridebat ;
Deinde fortunæ favore suffulti
Solum natale repetebant.

III.

Superioribus itidem sæculis
Magno animi robore,
Virorum multitudo
Ad Tamesis ripas direxit navem.

Alacritèr non sine insigni fortitudine
Milvum Odini sanguine pascebant *;
Ulterius porrò progressi
Magnis Angliæ Regibus militabant.

IV.

Larga sæpe munera acceperunt †
Strenui isti bellatores;
Soli pro carminibus præmii loco
Magnis cumulati divitiis ‡.

* Milvus Odini Corvum denotat, qui Odino sacer et admodùm familiaris fuit, unde et ipse Corvorum Deus in Eddâ et antiquâ Poesi dicitur.

† Sic, inter complures alios, *Egillus Skallagrimi* filius Islandus, insignis bellator *Adalsteni* Angliæ Regis castra secutus, fratrem suum *Thorulfum* in prælio quodam ibidem amisit, unde, satisfactionis et stipendii loco, binas arcas àrgento repletas a Rege accepit ad patrem deferendas, quas in Islandiam redux ipse retinuit, tandemque senex et luminibus sub mortem orbus, in palude quâdam submergebat. Occisis ibidem, qui ipsum manu duxerunt, duobus servulis. Prolixa hujus Athletæ vita, plurima etiam exotica hoc temporis tractu gesta continens, Islandorum manibus hodiedum teritur.

‡ Poetæ fuerunt Islandi ab initio optimi, unde in exterorum Regum et principum aulis, in summo semper honore et luce versati sunt, pro Encomiasticis, quæ haud raró tantis

Sedebant ultra mare
Compotationibus intenti,
Deinde honoribus aucti
Domum navibus remeabant.

V.

Sic etiam Anglorum naves appulere
Prisco tempore ad oras nostras,
Pretiosis vestibus et victu oneratæ,
Quod nos lucrum diu recordamur.
Vestes optimas lintea et funes
Nobis subministrârunt, usibus inservientia;
Nunc Islandis negatum est
His bonis diutiùs frui *.

nominibus obtulerunt, carminibus larga sæpenumero munera accipientes. Inter alios *Thorarinus Lostunga, Laudans lingua* appellatus, pro cantilenâ in honorem Canuti Magni Daniæ et Angliæ Regis confectâ, quam *Hösudlausn,* sive Capitis redemptionem vocavit, quinquaginta marcas argenti puri, muneris loco, recepit, testante Knutidarum vitâ.

* Angli seculo quatuordecimo et sequentibus insignia cum Islandis commercia exercuere, optimas semper nec unquam adulterinas merces advehantes, unde haud rarò eorum mercatores in Islandiâ hyemârunt, domibus passìm in hos usus exstructis. Ab iis qui ex Cambriâ huc appulerunt, loca quædam denominata sunt, ut *Kumbravogur* et sic porro. Hæc Anglorum commercia per varia interdicta in desuetudinem sensìm abierunt, jamque penitùs exspirârunt.

VI.

Hùc jam bonæ frugis viri,
Ab occiduis oris vela dirigere
Non detrectant. Angli nimirùm,
In sinu Hafnarfiord subsistentes,
Plurimis cluentem artibus
Peritum itineris antesignanum
Opum datorem omnes fatentur
Unum Josephum laude præditum.

VII.

Summâ profectò cum laude
Dominus BANKS nil cunctatus est
Per vastos pelagi fluctus
Navem dirigere;
Gloriosus suo cum comitatu
Per tres annorum orbes,
Vero cum honore perlustravit
Varias mundi plagas*.

VIII.

Ignotas hoc pacto detexit
Insulas, id quod assevero,
In vastâ et vorticosâ maris serie,
Optimas† plurimâ soli fertilitate;

* Profectus quippe Anno 1768, rediit in Angliam Anno 1771.

† Mari scilicet pacifico, quod non immeritò dixeris, cùm à Japponiâ ad Mexico MD mill: perhibeantur.

Harum una Otaheite, quam novimus,
Summâ amænitate conspicuam;
Ignorat imperium hyemis,
Omnigenâ felicitate circumdata*.

IX.

Lætus, per æquoreas undas
Navem aquilonem versus
Ventorum flatu celeritèr actam,
Nunc propellit pretiosarum vestium dator†,
Desiderans terram sub zonâ frigidâ‡
Perlustrare (nemo id prohibeat!).
Dignatur, id quod admodùm miror,
Pauperrimam gentem invisere.

X.

Alterum Solander asserimus,
Supremo illi honore proximum;
Hic optimis artibus excultus,
Ipsi Linnæo vix inferior§.

* *Otaheite* insula inter Americam et Asiam sita, ab armigero Banks detecta, optimæ et benignissimæ naturæ, atque proinde omnigenâ felicitate abundans. Invenit præterea *Novæ Zelandiæ* partem, in mari pacifico itidem sitam.

† Epithetum poeticum viri largi et opulenti, quod in Armigerum Banks optimè quadrat.

‡ Quædam enim Islandiæ pars jacet sub zonâ frigidâ.

§ Illi nempe decantatissimo rerum naturalium scrutatori

Indolem rerum investigat;
Plerosque antecellens.
Elegantem naturæ mystam,
Eruditi proinde collaudant.

XI.

Ter trinos præterea deprehendimus,
Gentis decora, bonâ eruditione
Et artium disciplinâ,
Ut optimè convenit, instructos
Omnes pari elegantiâ,
Tam largo opum diribitori
Per terras passìm et maria
Faventes semper adhærent.

XII.

Quid causæ quòd huc direxistis vela
Artificiosè texta, aquilonem versus?
Quid iter jam confectum causatur?
Plusquam modica animi delectatio:
Herbas, lapides, oculi exposcunt:
Optimatum sapientiam alit
Sparsas per territoria conspicere
Ignitas Heklæ montis scorias.

XIII.

Lunæ et lucidi Solis
Conjunctiones stellarumque orbes

Contemplantur admodùm sapientèr
Nobiles viri, ut ars augeatur.
Geiserem præterea convenire gestiunt *,
Et, si quæ plura audiuntur,
Manuscripta antiquitatis monumenta
Conquirere student honesti viri, ut ritè intelligant.

XIV.

Impensas omnes magnifico sumptu
Sine cunctatione agere
Haud gravatur
Anglus ille Heros, prout expedit.
Distribuit largè, ut novimus,
Lachrymas Sororis Freieri aurum †,

* *Geiser,* Nom. propr: aquæ æstuantis in Toparchiâ Arnesensi, australis Islandiæ, miræ profectò et reconditæ indolis; hujus contemplationi Angli Nostri integrum fermè diem impenderunt. Nomen habet a verbo islandico *ad giosa* evomere, ebullire; aquas enim per intervalla in altum evomit.

† In Eddâ et antiquâ Poesi Islandicâ, a Freyâ Odini uxore plurima auri Epitheta deducta sunt; nam juxta veterem Mythologiam lachrymæ Freyæ in aurum convertebantur; Pari modo, quod locuti sunt *Jötna,* seu Gigantes, in aurum mutabatur, unde *Jötna mal, Jötna tal,* Gigantum sermo, Gigantum loquela, pro auro apud Poetas haud raro ponitur. Aurum autem largè distribuere, jure dicitur Armiger Banks; cùm non tantùm aureis sed et aliis pretiosis rebus Islandos donaverit.

Eximias artes liberales promovet
Vir munificus magnæque dexteritatis.

XV.

Ambabus ergo ulnis
Excipite nobiles viros;
Subministrate omnia ex animo,
Equos largè suppeditantes.
Monstrate viam, prout optimè nôstis;
Acceptissimos viros per terram deducite.
Musæ hilari animo
Canentes talia depromunt.

XVI.

Salvete! (sic fari lubet)
Ad nos venientes, illustres Domini,
Quibus fortuna favet,
Commendabiles sapientiæ luce,
Fortuna vobis aspiret;
Prosperè cedant omnia itinera;
Favor cæli et felicitas viris
Facem per gelidam terram præferant!

XVII.

Nigricantes formæ usum Oculorum
Fortibus Anglis non intercipiant!

Adspiret lucens sol!
Contingat itidem videre cursum Lunæ!
Lucidæ stellæ claro lumine splendeant!
Ut sciant omnia juxta institutum ordinem
De Siderum situ
Sub Polo Arctico *.

XVIII.

Recedite subitò ad mare,
Recedite subitò densæ nubes!
Pluvia, Caligo, pulverisque vis
Ita aufugiant, ut non appareant!
Fumi vehementia divitibus viris
In oculos nequaquàm irruat!
Nihil amplius molestiam facessat,
Nil amplius iter reddat impeditum!

XIX.

In supremum Heklæ montis cacumen
Vestrum iter expediat
Fortuna laudem paritura,
Prout usibus vestris optimè inservit!

* Præprimis luminis borealis indolem ejusque causas eruere, quod eo magis optandum, quo majore difficultate res ista laborat, cùm in tot sententiarum divortiis quid de phenomeni hujus naturâ certo statuendum sit, adhuc ignoretur.

Ignis subterranei tecta latibula
Eruendi via vobis pateat
Adustosque lapides jacentes
In vastâ terræ superficie;

XX.

Herbas pretiosas disparuisse novimus,
Pallidæ quippe autumno emarcuere;
Proinde hæ grato ornatu
Viros excipere nequeunt.
Honorem interim elegantibus exhibere
Virgulta norunt passìm obvia.
Sabula fluvii et quæ iis innatant
Iter beatorum collaudent!

XXI.

Negotia omnia in vestrum honorem
Prosperè vobis succedant!
Augescat artium ludus
In austro per vestrum iter!
Eloquentes viri cumulatum gaudium
Et utilitatem ex itinere reportent,
Cùm hinc tendit navis
Occidentem versus *!

* Ita nobis loqui liceat, venerandæ Antiquitatis exemplo, quæ Angliæ, Scotiæ, Hiberniæ, Orcadumque incolas, *Vest-*

Grates vobis agimus, Domini erecti animi,
Grates vobis haud cunctantèr agimus
Prò honore nobis præstito;
Qui nobis imprimis gratus est.
Oras nostras nunquam inviserunt
Britanni, ut memoriæ proditum est,
Pari in Universum eloquentiâ præditi
Parique eruditionis laude inclyti.

XXII.

Salvo et incolumi curru
Domum hinc revehamini;
Viros quippe Anglos
Angelorum læta cohors deducat;
Rata maneant vota nostra,
Et fausta quæ ominamur
Vobis ex intimo corde,
Optimates artibus instructi!

XXIII.

Hinc per vastum æquor
Cùm navis celeri cursu tetenderit,
Secundi venti vela impleant!
Donec in occidente gradum sistit,

mannos, viros occidentales, eorumque terras *Vesterlönd*, terras occidentales, appellàrunt. Vid. inter alia *Landnam.* part. 1. cap. 5 et 7.

Et in vadis Anglici soli
Anchora jacitur,
Ut lætitiâ tandem perfusi
Exscensione factâ itineris labore levemini.

XXIV.

Nunquam prioribus seculis
Alii ex his mundi partibus
Heroicæ adeò indolis viri
(Naves) applicârunt ad nostros clivos *
Josephe, Te semper prædicat,
Te, Josephe, collaudat Islandia;
SOLANDER, decore notus,
Per Cygnorum † habitacula splendeat ‡

* Elogium hoc Angli nostri optimo jure promeruerunt; præ omnibus quippe exteris nationibus, oras nostras invisentibus, erga gentem Islandicam adeo munificos affabiles et humanos se exhibuerunt, ut simile vix reperire liceat exemplum. Eorum proinde adventum, itineris rationem, Nomina et raram benevolentiam, sempiternæ posteritatis memoriæ, annaliumque monumentis, grata me suasore conservabit Islandia.

† Territoriam Alstanes ad sinum Hafnafiord, ubi navis Anglorum in anchoris consistebat, a cygnis nomen habet, unde tractum hunc *Svanabigder*, seu Cygnorum habitacula, nominamus, habito simul ad universam insulam respectu.

‡ Splendeat cluentis famæ præconio.

XXV.

Valete, humanissimi Domini;
Laus vestra super terram
Per longam vivat ætatem!
Laudent vos quæcunque moventur!
Plura loqui supersedemus
Doctæ Sorores gratiam sitientes *:
Hoc gratulatorium Carmen non ingratum
Londini et Assatuni † palàm pronunciamus.

N° 4.

VIRIS

ILLUSTRISSIMIS, NOBILISSIMIS, SUMME REVERENDIS, AMPLISSIMIS ET CONSULTISSIMIS!

PLURIMA SALUS!

Vestra hìc in Schalholtiâ conversatio omnibus erat grata, quare in omnium ore versamini in celebritate, immò indigenarum laude et memoriâ.

* Gratiam tantorum virorum sibi humillimé apprecantes.

† *Assatun* antiquitùs urbs Angliæ; forte non procul a Brandfurdâ, hodie Brentfort, ad quam utramque Canutus magnus, cum filiis Adalradi Angliæ Regis conflixit; sed in quâ parte Regni sita fuerit, et num hodie sub alio nomine existat, juxta ignoramus. Sed hìc metri tantùm causâ adhibetur.

Verùm post vestrum discessum res novæ mihi sunt relatæ, quòd totum orbem intra triennium emensi fueritis; quo vasto itinere audito, miratio mihi facta est, immò omnes in admirationem vestri trahit, suo animo apud se perpendentes, quòd non minùs incredibili animi robore septi, quàm firmissimâ corporum complexione præditi sitis. Quod Jacobus de Lamaire et Wilhelmus Soutensis terrarum orbem peregraverint, memoriæ est proditum: vestrum susceptum molimenque omnium est cultu dignum. Præterea liberalitas vestra penè regifica omnium meritò retinet animos; qualis munificentia rara est in exemplis, Præsulis Mag. Johannis Widalini exceptâ, cujus super anniversaria Evangelia orationes incolæ habent, a Te, Nobilissime Troili, Schalholti emptas. Huic olìm objectum, parsimonium optimum esse vectigal, seram esse in fundo, opus esse cauto: ad quod Præsul; Mors semper impendet, nunquam longè abesse potest. Accedit stabilis atque ingenuus mos et decor, ac inoffensa gravitas, nec sine jucunditate senile illud vix dum viriles annos ingressi pondus, illaque exacta frugalitatis lex, sine quâ frustrà aliquis mentem applicet magnis; modestiaque inusitata, quæ neque summis, ut ille ait, mortalium spernenda est, atque a Diis æstimatur. Igitur cùm vos ipsos loco ornatissimos lepidis exornetis moribus, singularibusque studiis, tùm meum adjicere decet suffragium, et unà cum aliis laudibus exquisitis ornare. Quod reliquum est, Poetæ verbis,

> Sic te Diva potens Cypri,
> Sic fratres Helenæ, lucida sidera,
> Ventorumque regat pater,
> Obstrictis aliis, præter Iapyga!

Vos, illustres optimos, saluto! Quid dico, poetæ et ipsâ Apostolicâ salutatione amplector; vos Deo Opt. Max. cujus nutu et arbitrio omnia reguntur, omniaque vestra commendare non desistam. Ipsissimus vestrum per orbem iter fortunet, et ad umbilicum ducat! postremò patriis redditi terris, ad summas in cælesti curiâ evehat dignitates.

Hæc pauca boni æquique consulere dignemini.

Vestræ nobilitatis

Addictissimus,

THEODORUS JOHANNIS, P. Em.

Schalholti, Isl. d. 27, Dec. 1772.

P.S. In hoc vobis, viri nobilissimi, gratulor quòd igniflua, ignivoma, et crudelis Virgo (Hekla) cum pace incolumes dimiserit.

Viris perillustribus et excellentissimis,
Magistro Josepho Banksio,
Doctori Danieli Solandro, et
Magistro Troilio.

HYRDUGLEGUM HERRA	HONORABILI DOMINO
Alexander Jones,	*Alexand. Jones,*
Hofudsmanni hins Stor-Brettendska	Navis Bellicæ Reg. Britann.
STRIDSKIPS	THE TALBOT,
THE TALBOT.	PRÆFECTO.
Huldi myrkur mockur villu menta sol mengi Kristnu.	Obscurus errorum vapor solem scientiarum po- pulis Christianis pertexit.
Bio Pafa flœrd prisund salum kryndir tirannar kroppum hlecki.	Astutia Paparum ani- mis custodias, tyran- ni coronati corporibus vincula instruxerunt.
Bonnudu dolgar boknam leikum og modur mals medferd rietta;	Prohibuerunt nebulones isti secularibus littera- rum studia, et linguæ vernaculæ rectam tracta- tionem.
Tha nysjen i Nordur hafi kynntiz farendum soldardili.	Tum in mari septentrio- nali, nuper visa terræ particula navigatoribus innotuit.

Fadm baud eya isum vafinn neyddri thjod ur nanum londum.	Sinum præbuit insula, glacie circumdata, genti- bus oppressis e vicinis regionibus ;
Syrrtuz thar strialsir holdar grœdgi muks og grams ofbeldi,	Effugerunt hìc viri liberi voracitatem clerico- rum et violentiam principum :
Austmenn, Danir, Irar, Bretar, bygdu folld og blomgvaz letu	Norvegi, Dani, Hiber- ni, Britanni, terram fre- quentârunt et floresce- re fecerunt.
Fjeck theim frelsi frœdi skoput; Rit og mal rietti nadu.	Libertas artes pa- ruit ; sermo et scriptura jus suum obtinuerunt.
Gall thar greppa gullinn harpa fedra saung frœg threkvyrki.	Sonuit ibi poetarum au- rea Lyra ; patrum ce- cinit celebria faci- nora.
Sin og kunnra samlifendra	Sua et notorum coæ- vorum fata et acta

Orlog ok verk
adrir Skradu;

Skalda log
Skrifs og mœlsku
lista og ydna
lietu ritinn.

Undrast enn
Europear
frodir visindi
fedra vorra.

Breyttiz fron
breyttuz landar,
œrduz their auds
og œru-syki.

Mottu hvartveggia
mentum fremur;—
Kongs og klerka
kugan hrepptu.

Soadi mengi
svartur Daudi *

alii composue-
runt;

Leges canendi, scriben-
di, dicendi, artium
et opificiorum scrip-
tis tradiderunt.

Mirantur adhuc
docti Europæi
atavorum nostrorum
eruditionem.

Mutabatur terra,
mutati sunt incolæ,
divitiarum et gloriæ a-
viditas infatuavit istos.

Utraque scientiis præ-
posuerunt;—Regis
et cleri oppressionem
nacti sunt.

Mors atra * populum
devoravit;—laus

* Svanefnd drepsott.

* Pestis ita nominata.

hvarfur landi hrodur forni.	antiqua emigravit e terrâ;
Foruz listir fje og sæla vesladiz fold og fegurd tindi :—	Perierunt artes, opes et felicitas ;—solum de- terioratum amænita- tem amisit.
Enn i Nordalfu onnur riki sœld og visindi sunnann fluttu.	Sed in Europæ cætera regna scientiæ et beati- tudo e meridie trans- migrârunt.
Hvurfu svo morg hundrud ara, œ Gardarsholma haignandi for;	Diffluxerunt ita anno- rum aliquot centuriæ; status Gardaris insulæ assiduò decrevit;
Uns mildingar mœrir Dana vyrdtuz hans mein vilja bœta,	Quousque clementes Da- niæ Reges damna istius reparare desiderabant;
Leystu verdslunar vonda hlecki gafu gots til groda fyrdum,	Solventes noxias commer- cii catenas, pecuniam offerentes ad rem in- colarum augendam.

Elldur, is,	Ignis, glacies, tabes,
orœkt jardar,	terræ increta (nostratum)
orbyrgd mognud	egestas, proposito genero-
moti stodu,	so obstiterunt ;
Enn orbyrgd tha	inopiam autem istam, An-
engill fridar	gelus pacis, animo sort
gjœddi anœgju	suâ læto, auro præstar
gulli betri.	re, subdidit.
Sast ej dreginn	Nullibi videbatur gla-
dor a lopti	dius vibratus, nec tor-
nje fallbyssur	menta bellica, perni-
fylltar heli	cie plena ;
Ej heiptug hond	nec manus hostilis san-
hellti blodi,	guinem effundens, nec-
nje herlogar	flammæ classicæ domi-
hreysi gleyptu.	cilia vorantes.
Saum Isalands	Vidimus procellosum
einu sœlu	turbinem solam Is-
hvirfilbil	landiæ prosperitatem
hradt kollverpa,	humi rapidè prosternare.
Saum frid vorn	Vidimus pacem nostram
og fedra vorra	et patrum nostrorum tan-

loks far flotta	dem pallescentem in fu-
folvann hverfa.	gam verti.
Allur umhverfis	Totus circumcirca con-
œstiz heimur	turbabatur orbis bel-
bali blodgu	li cruentis incendiis ;
brandasennu ;	
Drundu skjœdar	Perstrepuerunt univer-
Skruggur viga	sum mare hominibus
of sœ allann	noxia cædium tonitrua.
at seggia mordum.	
Tho i fjarska leit	Eminùs tamen aliquam-
thessi undur	diu spectabat prodigia
vor um hrid	ista extrema hæcce mundi
heims utkjalki	regio.
Mattkum budlungi	Potenti Regi Magnæ
Bretlands mikla,	Britanniæ, Dano-
ovin vordnum	rum omnium hosti facto,
allra Dana,	
	Visum fuit Islandiæ
leitst Islandi	miseræ parcere, nolenti e-
aumu hlifa	am suarum classium in-
fyrir arasum	cursa-
flota sinna ;	tionibus infestatam fore ;

Enn hans thegna
theinktu nockrir
gull i gull-lausu
grœda landi.

Ipsius autem subditorum
quidam, aurum sperabant
se in terrâ auro destitutâ
acquisituros.

Hleypti Gilpingr
a herfleyi
Vikur til
at veidum aura.

Adduxit Gilpinus Rei-
kiavicæ navem armatam,
ad captandas divitias;

Hlaut hann ej Skjell
fyn skillinga
enn svipti fron
fjarsjod stokum,

Non pro nummis verbera
nactus, abstulit hìnc
provinciæ unicum æ-
rarium,

Theim er atti
Thurflir bœta
fatœks lyds
og logfrid verja.

Auxilio pauperum et præ-
sidio securitatis publicæ
destinatum.

Adrir baru
at Kaupstefnu
vopn og thannig
varning byltu.

Nundinas alii armigeri
frequentabant, merces
ita permutantes.

Lands tho log
lyda vorra

Pacem tamen internam
gentis nostræ leges pa-

innbyrdis grid alljafnt vordu.	triæ adhuc defenderant.
Saum loks a sumri thessu gior umrotad ro almennri;	Vidimus tandem, hâc æstate, publicam quie- tem penitùs disturba- tam;
Vorum frœmsta folkstyranda byllt ur ondvegi ordnum fanga.	Primarium provinciæ Procuratorem solio de- jectum et in carcerem detrusum.
Upphlaups litum andann ljota geysa of allt i alvœpni.	Vidimus seditionis hor- ribilem dæmonem, ar- mis succinctum, om- nia obruere.
Ljetst hann Engla lofdung thiona hermaktar hans hafa fylgi,	Simulavit se Anglorum Regi servire, istius exer- cituum favore nisum.
Vopnadir brodir bormum moti enn otti greip adra lydi,	Frater se in fratres armavit;—rapuit error reliquum popu- lum,

Hofdu ej sjed sverd nje dreyra, lagaleysi lutu naudgir.	qui nunquam ante en- sem aut sanguinem con- spexerat, et invitus col- la jugo insolentiæ sub- didit.
Sa hinn oblgari ebldi vyrki og heldocku hreykti merki.	Fortior ille munimen- ta ponens, orci instar, atrum vexillum erexit.
Tok hann tignar titil jarla, vogandi mildings makt sier eigna.	Ducalem sumpsit digni- tatem, sibi Regiam po- testatem arrogare ausus.
Ljest af thjod vorri Thar til knuinn at hun uppreystar oll svo krefdi;	Prætexuit gentem nos- tram hæc a se deprecatam fuis- se, ipsamque tales turbas un- animen poscere;
Frid og frelsi sig fœra sagdi theim, er vafdi thrœla-hleckium.	Dicens se pacem et liber- tatem adferre ipsis iis- dem, quos servilibus onera- bat catenis.

Ottuduz their tr annad vildu eru og mannœra tinaz mundu;	Metuebant contraria optantes fidem et honestatem omninò abolitam fore;
Ad landradum lucka styrdi, og osigrad Englands merki.	perduellioni fortunam necnon Angliæ invictum vexillum favere ominantes.
Thannig harmendra huggun besta, himnesk von var oss fluinn.	Sic mœstorum optimum solamen, cælestis spes, aufugit.
Sem langhrakinn Sœfarandi, barinn skjelfingum brims og storma,	Ut navita, per longum temporis spatium, æquoris et tempestatis impetibus vexatus,
er i skjœdum Skerjagardi, sier vid Skrugguljos, Skip sitt sveyma,	qui, fulgure lumen præbente, navem inter scopulos insidiosos jactari nupèr conspexit,
Enn hofn engva hrjadu fleyi,	portum autem nullum,

nema afgrium eitt hndirdjupa—	nisi in medio oceani abysso,
Threlsazur fari thessu sem med ovœntu Undraverki ;	jam ex his eripitur pe- riculis, ceu numinis nutu ;
Verdur blœlogn ur bilvedri, slatur sœr ur fjallbylgjum,	vertitur in tranquilli- tatem procella, mare undis montuosum in malaciam,
Solbyrta ur blisum reidar,— leidur hœgur byr ad hofnum fley ;—	in solarem splendorem fulminantia fulgura— aura prospera et pla- cida portum versus navi- gium ducit :—
Thannig fœrdi thin adkoma, edallundadi ALEXANDER ! fognud og frelsi froni pessu, sviptur thu thad ognum ostjornunar.	Sic tuus adventus, generose ALEXANDER ! gaudium et libertatem ter- ræ huic attulit ; levasti istam Anarchiæ horro- ribus.

Hvar mun syna sega veraldar veitta af fjendum veljord slika?	An tantum benefi- cium, ab hoste hosti datum, monstret historia mundi?
veit eg hana hœla thinum fœgsta nafna fyri minna.	Certus scio celeberrimum tuum cognominem le- viore magnanimitate perennes laudationes acquisivisse.
Ej fyri makt nje mauradyngjur metord nje hros hjalp oss greiddir:	Non dominationis, nec divitiarum, nec digni- tatis, nec laudis gratiâ auxilium nobis præbuisti.
Idgjold onnur, eru likindi bjodiz their, enn thess, thu ej qvidir.	Alia præmia, ut videtur, tibi afferentur, quæ tamen non metuis.
Leyf ad thidar thier thackir greidum, og formonnum odrum fylgdar thinnaz;	Sine ut tibi et re- liquis tui ordinis ductoribus, singula- res agamus gratias!

Ey skal minning
ydar deya
i bokmenta
bolstad forna

Nunquam vestra abo-
lebitur memoria in
prisco hocce musarum
habitaculo.

Thaklat er ond
Offri kjœrri
theim annars ej
Offurs krefur.

Sacrificio potiùs pla-
cet anima grata isti,
qui victimam petere
nolet.

Leyf, thig enn
lids at bidjum,—
Oss frarœndum
folk-narungi

Sine etiam, ut ulte-
rius a te auxilium pe-
tamus—quòd nempe
primati, nobis vi ademp-
to,

Thier samedla,
ad sid og œttum
hialp, sem fyrr,
holla veitir,

tibi et indolis et stir-
pis nobilitate pari,
adminiculum, ut ad-
huc, porrò etiam præ-
beas,

riett ad sok
sigur vinni
og fullkomnun
fro vor nai.—

ad bonæ causæ tri-
umphum, nostrique
perfectionem gaudii
suo tempore compa-
randum.

Oskum vœr thier	Precamur tibi et tuis
og thinum lydum,	honorem, salutem et
heidurs, audnu,	lætitiam, nobisque
og anægju,	junctìm annonæ
enn oss ollum	et pacis felicitatem !
ars og fridar !	

Mense Augusto, Anni 1809.—*in Islandia.*

END OF APPENDIX. D.

APPENDIX. E.

ICELANDIC PLANTS.

APPENDIX. E.

LIST*

OF

ICELANDIC PLANTS.

I. MONANDRIA.

I. MONOGYNIA.

HIPPURIS vulgaris.
Zostera marina.

* This catalogue is principally taken from Zoega's *Flora Islandica* (attached to the Danish edition of Povelsen and Olafsen's account of Iceland), and Mohr's *Forfög til en Islandsk Naturhistorie,* published at Copenhagen in 1786. The few additional species, which I am enabled to insert by means of Sir George Mackenzie's and Mr. Paulsen's collections and my own researches, are distinguished by being printed in italics. In some instances, where I have, in the course of my journal, had occasion to notice any new plant or any peculiarity belonging to those that are already known, I have referred to the page where it is mentioned.

II. DIGYNIA.

Callitriche aquatica.
——— ——— γ. autumnalis.

II. DIANDRIA.

I. MONOGYNIA.

Veronica officinalis.
——— serpyllifolia.
——— Beccabunga.
——— Anagallis.
——— scutellata.
——— alpina.
——— fruticulosa. (vol. I. p. 113.)
——— marilandica.
Pinguicula vulgaris.—" Les Islandais s'en servent quelquefois en guise d'ail." *Voyage en Islande.*
——— alpina.

II. DIGYNIA.

Anthoxanthum odoratum.

III. TRIANDRIA.

I. MONOGYNIA.

Valeriana officinalis.
Schœnus compressus.

Scirpus palustris.
——— lacustris.
——— cæspitosus.
——— acicularis.
——— setaceus.
Eriophorum polystachion.—Of the *pappus* of this plant the natives make wicks for their lamps.
——— vaginatum.
——— capitatum. *Hoppe*. (vol. I. p. 178.)
——— alpinum.
Nardus stricta.

II. DIGYNIA.

Phleum pratense.
——— nodosum.
——— alpinum.
Alopecurus geniculatus.
Milium effusum.
Agrostis rubra.
——— stolonifera.
——— canina.
——— vulgaris.
——— ——— β. pumila.
——— alba.
——— arundinacea.
——— cærulea.

Aira cæspitosa.
—— flexuosa.
—— montana.
—— subspicata.
—— alpina.
—— aquatica.
—— præcox.

Holcus odoratus.—Said to be used by the Icelanders to perfume their apartments and their clothes.

Sesleria cærulea.

Poa pratensis.
— trivialis.
— compressa.
— annua.
— angustifolia.
— alpina.
— maritima.
— *glauca.*—Both this and the following species are far from uncommon in Iceland.
— *cæsia.*

Festuca ovina.
——— rubra.
——— elatior.
——— fluitans.
——— duriuscula.

Festuca *vivipara.* (vol. I. p. 320.)
Arundo Phragmites.
——— Epigejos.
——— arenaria.
Elymus arenarius.—(vol. II. p. 226.) The seeds are occasionally made into a sort of bread.
Triticum caninum.
——— repens.

III. TRYGINA.

Montia fontana.
Koenigia islandica. (vol. I. p. 152 and 191.)

IV. TETRANDRIA.

I. MONOGYNIA.

Scabiosa succisa.—The Icelandic names for this plant, *Pukabit* and *Dievelsbid,* have both the same signification as our *Devil's bit.*
Galium verum.
——— palustre.
——— Mollugo.
——— *pusillum.*
——— boreale.
Plantago major.

Plantago lanceolata.
——— maritima.
——— *alpina.*—This I recollect seeing, in some plenty, at Thingevalle, and I have since received specimens from Sir George Mackenzie and Mr. Paulsen.
——— Coronopus.
Sanguisorba officinalis.
Alchemilla vulgaris.
——— alpina.

III. TETRAGYNIA.

Potamogeton natans.
——— marinum.
——— compressum.
——— lucens.
——— crispum.
——— perfoliatum.
——— pectinatum.
——— pusillum.
Sagina procumbens.
Tillæa aquatica.

V. PENTANDRIA.

I. MONOGYNIA.

Myosotis scorpioides. α and β.
Pulmonaria maritima.

Echium vulgare.
Primula farinosa.
Menyanthes trifoliata.—This plant is important to travellers who are not acquainted with the route in the morasses; for they are well aware that wheresoever it grows they may safely pass; its closely woven roots making a firm bed upon the soft subsoil. The Icelanders call it *Reidinga,* and employ the matted tufts to prevent the saddle or any load from chafing the horses' backs.
Azalea procumbens.
Campanula rotundifolia.
——— patula.
Viola canina.
—— tricolor.
—— palustris.
Glaux maritima.

II. DIGYNIA

Gentiana campestris.
——— amarella.
——— nivalis.

Gentiana aurea.
——— detonsa.
——— bavarica.
——— tenella.—(filiformis of *Persoon's Synopsis.)*
——— verna.
——— rotata.
Hydrocotyle vulgaris.
Ligusticum scoticum. (vol. I. p. 323.)—To this plant, which Mr. Paulsen named by mistake *Imperatoria Ostruthium,* was attached the following observation: "Hæc (in Islandiâ) rarissima herba, in saxis solùm et montibus præruptis maritimis reperiunda. Devoratis radicibus hìc traditur divinos edidisse effectus in hydaridibus abdominalibus. (isl. *meinlœti*)."
Angelica Archangelica. (vol. I. p. 191.)—The Icelanders gather the stems and roots of this plant, which they eat raw, and generally with the addition of fresh butter.
——— sylvestris?
Imperatoria Ostruthium.

Carum Carui.—Naturalised in Iceland, according to Sir George Mackenzie.

IV TETRAGYNIA.

Parnassia palustris.

V. PENTAGYNIA.

Statice Armeria.
Linum catharticum.

VI. HEXAGYNIA.

Drosera rotundifolia.
——— longifolia.

VI. HEXANDRIA.

I. MONOGYNIA.

Convallaria biflora.
Juncus effusus.
——— *arcticus*.—Discovered by Sir George Mackenzie.
——— squarrosus.
——— trifidus.
——— articulatus.
——— bulbosus.
——— bufonius.
——— biglumis.
——— triglumis.

Juncus pilosus.
——— campestris.
——— spicatus.

II. TRIGYNIA.

Rumex digynus.—All the species of *Rumex* are boiled and eaten by the Icelanders; though only the young shoots of *acutus* are employed. Of the *Acetosa* a beverage is made by the common people, by steeping the plant in water till all the juice is extracted. This drink is kept some time; but soon becomes bad and putrid in warm weather.
——— acutus.
——— Acetosa.
——— Acetosella.
Triglochin palustre.
Triglochin maritimum.
Tofieldia palustris.

VIII. OCTANDRIA.

I. MONOGYNIA.

Chamænerium halamifolium (vol. I. p. 260 and 318.)—From specimens now before me it appears that this species is subject to considerable variation, as well in the proportional breadth of its leaves, as in the size of the flowers. Mr. Paulsen remarks "Crescit ferè solum ad fluvios montium glacialium, in argillâ et arenâ vulcanicâ."

Chamænerium angustifolium. (vol. I. p. 322.)

Epilobium montanum.

——— palustre.

——— *origanifolium.*

——— alpinum.

——— tetragonum.

Vaccinium Myrtillus.

——— Oxycoccos.

——— uliginosum.—For its uses see vol. I. p. 215.

Erica vulgaris.—"Ex ejus magnâ florescentiâ de magnâ nivis hyemalis copiâ augurantur Islandi." *Paulsen in Epist.*

II. TRIGYNIA.

Polygonum viviparum. (vol. I. p. 113.)

Polygonum Bistorta.—The roots are often eaten raw, and sometimes converted into bread.
——— Hydropiper.
——— amphibium.
——— Persicaria.
——— aviculare.

III. TETRAGYNIA.

Paris quadrifolia.

X. DECANDRIA.

I. MONOGYNIA.

Andromeda hypnoides. (vol. I. p. 85 & 122.)
Arbutus Uva Ursi*.–See note in vol. 1. p. 217.
——— alpina.
Pyrola rotundifolia.
—— secunda.
—— *minor*. (vol. I. p. 122.)

II. DIGYNIA.

Saxifraga Cotyledon.
——— stellaris.

* "The leaves of this are used in Lapland for tanning and dyeing, which saves a great deal of alum. Many barrels of these leaves are sent for sale to Stockholm." *Linnæus' Lachesis Lapponica*, v. I. p. 250.

Saxifraga nivalis.
——— Hirculus. (vol. I. p. 254.)
——— *palmata.*
——— punctata.
——— oppositifolia.
——— autumnalis.
——— aizoides.
——— bulbifera.
——— *cernua.* (vol. I. p. 209.)
——— rivularis. (vol. I. p. 116.)
——— tridactylites.
——— cæspitosa.
——— groenlandica.—My specimens of this, gathered by Sir George Mackenzie, exactly accord with the figure of this species in the *Flore des Pyrenees.* La Peyrouse has observed it growing at the height of sixteen hundred toises above the level of the sea.
——— hypnoides.
——— *tricuspidata.*
——— petræa.
Scleranthus annuus.

III. TRIGYNIA.

Silene maritima.

Silene acaulis.—Boiled and eaten with butter by the Icelanders.

Stellaria media.

——— biflora.

——— cerastoides.

Arenaria peploides.—This is steeped in sour whey, where it ferments: then the liquid is strained off, and fresh water added to the beverage, which is said to taste like olive-oil; whence the name of the plant in Iceland, *Smidiukaal.—Voyage en Islande.*

——— serpyllifolia.

——— ciliata.

IV. PENTAGYNIA.

Sedum *saxatile.*

——— rupestre.

——— annuum.

——— acre.—" Vulgatum in Islandiâ vomitorium."—*Paulsen in Epist.*

——— villosum.

Lychnis Flos-Cuculi.

——— alpina.

——— ——— *var. fl. albo.*

Cerastium viscosum.

——— vulgatum.

Cerastium alpinum.
——— *latifolium.*
Spergula arvensis.
——— nodosa.
——— saginoides.

XII. ICOSANDRIA.

II. PENTAGYNIA.

Pyrus *domestica.*—This was found by Sir George Mackenzie, growing eight feet high, in a cleft of lava near Budenstad in Snöefel-syssel. Another plant of the same was also discovered by the same gentleman at Eyafiord, on the north coast.
—— aucuparia.
Spiræa Ulmaria.

III. POLYGYNIA.

Rosa hibernica.—This, the only species of *Rosa* discovered in Iceland, was sent me by Mr. Paulsen, with the following remark: "Nulli hìc priùs obvia. Crescit in rupe unicâ ad villam Seljaland."
Rubus saxatilis.

Fragaria vesca. (vol. I. p. 83.)
Potentilla verna.
——— anserina.—The roots are frequently eaten in the southern parts of the island.
——— aurea.
Tormentilla officinalis.—In Iceland, I am not aware that any use is made of this plant; although in Lapland, according to Linnæus, it is chewed along with the inner bark of the Alder, and the saliva thus impregnated is applied to leather to dye it of a red color.
Geum rivale. (vol. I. p. 235 and 268.)
Dryas octopetala.—Its leaves, as observed in vol. I. p. 47, are gathered, and made into a sort of tea.
Comarum palustre.

XIII. POLYANDRIA.

I. MONOGYNIA.

Papaver nudicaule. (vol. I. p. 323.)

V. POLYGYNIA.

Thalictrum alpinum.

Ranunculus acris.—Often used for making blisters.

——— hederaceus.

——— reptans.

——— aquatilis.

——— lapponicus (vol. I. p. 191.)

——— repens.

——— glacialis.—A rare plant in Iceland. I was not so fortunate as to meet with it myself. Sir George Mackenzie has favored me with the only specimen which he procured: it was found growing among loose stones on the declivity of a mountain between Stadarhraun and Kolbein-stadr.

——— nivalis.

——— hyperboreus.

Caltha palustris.

XIV. DIDYNAMIA.

I. GYMNOSPERMIA.

Lamium purpureum.

Galeopsis Ladanum.

——— Tetrahit.

Thymus Serpyllum.—An infusion of the leaves is often used to give an aromatic flavor to the sour whey.

Prunella vulgaris.

II. ANGIOSPERMIA.

Bartsia alpina. (vol. I. p. 270.)

Rhinanthus Crista-Galli.

Euphrasia officinalis.—I possess alpine varieties of this plant from Iceland, which (though bearing perfect flowers) scarcely rise a quarter of an inch above the surface of the ground.

Pedicularis sylvatica.

——— flammea.

Limosella aquatica.

XV. TETRADYNAMIA.

I. SILICULOSA.

Subularia aquatica.

Draba verna.

—— muralis.

—— incana.

—— ——— *var. contorta.* ***Retzius.***

Thlaspi Bursa Pastoris.

——— campestre.

Cochlearia officinalis.

——— danica.—Occasionally eaten as spinage, and reckoned of service in the cure of the scurvy, though seldom made use of.

Bunias Cakile.

II. SILIQUOSA.

Cardamine pratensis.

——— hirsuta.—A singular variety of this plant, if not a distinct species, has been sent me both by Sir George Mackenzie and Mr. Paulsen, having the lower leaflets round, the upper ones linear, and all very entire.

——— bellidifolia.

Sisymbrium terrestre.

Arabis alpina. (vol. I. p. 268.)

—— hispida.

Brassica alpina.—Sent me by Sir George Mackenzie.

XVI. MONADELPHIA.

V. DECANDRIA.

Geranium sylvaticum.

——— pratense.

——— montanum.

XVII. DIADELPHIA.

III. OCTANDRIA.

Polygala vulgaris.

IV. DECANDRIA.

Lathyrus pratensis.
Vicia cracca.
Pisum maritimum.
Lotus corniculatus.
Anthyllis vulneraria.
Trifolium arvense.
——— pratense.
——— repens.—" Les gens de la campagne, dans la partie Nord et Est de cette ile, en mangent en légume."—*Voyage en Islande.*

XIX. SYNGENESIA.

I. POLYGAMIA ÆQUALIS.

Leontodon taraxacum.
——— autumnale.
Hedypnois Taraxaci.
Hieracium Pilosella.
——— Auricula.
——— alpinum.
——— præmorsum.

Hieracium Murorum.
Serratula arvensis.
Carduus lanceolatus.
——— heterophyllus.

II. POLYGAMIA SUPERFLUA.

Gnaphalium alpinum.
——— uliginosum.
——— *sylvaticum.*
——— fuscatum. *Pers.*
Erigeron alpinum.
Senecio vulgaris.
Pyrethium inodorum.
——— *maritimum.*
Achillea Millefolium.—The Icelandic appellation, *Vall-humall* (field-hops), seems to imply that this plant has been used instead of hops in that island, as it is still in some parts of Sweden. At present the natives only make an ointment of its leaves with butter, which they apply to cutaneous and other external sores.

XX. GYNANDRIA.

I. DIANDRIA.

Orchis maculata.

Orchis Morio

—— mascula.

—— latifolia.

—— hyperborea. (vol. I. p. 85.)

Satyrium viride.

——— albidum.

——— nigrum.

Epipactis ovata.—I possess the only specimen of this ever gathered in Iceland; it was found, at a place called *Vik*, by the son of Mr. Paulsen.

——— *Nidus avis?*—Either this or a new species of *Epipactis* has been sent me by Sir George Mackenzie. The specimen is destitute of its root, so that I cannot ascertain it with certainty.

Cymbidium Corallorhizon.

XXI. MONŒCIA.

I. MONANDRIA.

Zostera marina. (vol. I. p. 111)—This the cattle eat, and the natives gather and dry for their beds.

Chara vulgaris.

—— hispida.

III. TRIANDRIA.

Sparganium natans.

Cobresia scirpina. Willd. (vol. I. p. 152 and 178.)—*Carex Bellardi* of preceding authors, under which name it is mentioned in my journal.

Carex dioica.

—— capitata.

—— pulicaris.

—— arenaria.

—— uliginosa.

—— leporina.

—— vulpina.

—— muricata.

—— loliacea.

—— canescens.

—— elongata.

—— flava.

—— pedata.

—— montana.

—— rigida.

—— limosa.

—— *atrata.* (vol. I. p. 116.)

—— pallescens.

—— capillaris.

Carex Pseudo-cyperus.
—— acuta.
—— *ampullacea.*—The specimen sent me by Sir George Mackenzie is a slight variety with branched spikes.
—— vesicaria.
—— hirta.

IV. TETRANDRIA.

Urtica dioica.
—— *urens.*—This I only saw growing in Mr. Savigniac's garden, at Reikevig.

VIII. POLYANDRIA.

Myriophyllum spicatum.
———— verticillatum.
Ceratophyllum demersum.
Betula alba.
—— nana. (vol. I. p. 241.)

XXII. DIŒCIA.

II. DIANDRIA.

Salix Myrsinites.
—— arbuscula.
—— herbacea.—The downy substance from this and other species of Willow is applied by the natives to wounds

both of man and beast. The leaves steeped in water are employed in tanning skins. The wood is used in making ink, being steeped in a decoction of the leaves, to which is added some of the earth used in dying, mentioned in the note, at vol. I. p. 215: it is then all boiled together until the liquid has acquired a proper consistency.

Salix purpurea.
—— reticulata.
—— myrtilloides.
—— glauca.
—— lanata.
—— Lapponum.
—— arenaria.
—— fusca.
—— caprea.
—— pentandra.

III. TRIANDRIA.

Empetrum nigrum.

VIII. OCTANDRIA.

Rhodiola rosea.

XIII. MONADELPHIA.

Juniperus communis.

XXIII. POLYGAMIA.

I. MONŒCIA.

Atriplex laciniata.
——— patula.

XXIV. CRYPTOGAMIA.

I. FILICES.

Equisetum sylvaticum.—Various species of *Equisetum* are given to the cattle in Iceland, where they are said to be excellent food for the saddle-horses.
——— arvense.
——— limosum.
——— palustre.
——— fluviatile.
——— hyemale.
Osmunda Lunaria. (vol. I. p. 116.)
Ophioglossum vulgatum.
Lycopodium alpinum.—For its use in dying woollens, see vol. I. p. 214.
——— clavatum.
——— annotinum. (vol. I. p. 85.)
——— Selago.
——— selaginoides.

Lycopodium dubium *.
Polypodium vulgare.
——— fontanum.
——— ilvense.
——— arvonicum. (vol. I. p. 58.)
——— Phegopteris.
——— Dryopteris. (vol. I. p. 320.)
Aspidium Lonchitis.
——— Thelypteris.
——— Filix mas.
——— Filix fæmina.
——— fragile.—I possess a curious and elegant species of *Aspidium* (*Cyathea* of Doctor Smith) somewhat allied to this, but hitherto undescribed.
Asplenium septentrionale.
Isoetes lacustris. (vol. I. p. 208.)

II. MUSCI.

Phascum muticum.
Sphagnum obtusifolium.—The same use being made of this moss in

* Surculis simplicissimis, erectis, compressis; foliis complicatis, carinatis, acutis, alternis, distichè imbricatis. *König.*

Iceland as in Lapland, I shall be readily excused for inserting Linnæus' words upon the subject. "Feminis *Lapponicis* maxime notus est hic muscus; hunc enim, linteis cùm destituantur, infantibus, dum cunis suis continentur, undique circumponunt, qui et pulvinaris et tegmenti vices servat, urinam acrem absorbet, calorem conservat, sericisque stragulis gratior est tenellis; mutatur deinde vesperi et mane, dum purus et recens substituitur in prioris locum."—*Fl. Lapp. p.* 337.

Sphagnum capillifolium.
Gymnostomum truncatulum.
——————— *fasciculare.* (vol. I. p. 50.)
Tetraphis pellucida.
Andræa rupestris.
——— *Rothii.* (vol. I. p. 154.)
Splachnum ampullaceum. (vol. I. p. 88.)
———— urceolatum.
———— mnioides.
———— rubrum.

Splachnum vasculosum. (vol. I. p. 260.)
Conostomum boreale. (vol. I. p. 85 and 94.)
Encalypta vulgaris.
——— *alpina.* (vol. I. p. 83.)
Grimmia apocarpa.
——— *maritima.*—Not uncommon on rocks by the sea shores.
Weissia cirrata.
Encalypta *lanceolata?* (vol. I. p. 23.)
Dicranum scoparium.
——— undulatum.
——— heteromallum.
——— purpureum. (vol. I. p. 113.)
——— flexuosum.
——— squarrosum.
——— *pusillum.*
——— pulvinatum.
——— taxifolium.
——— hypnoides. (vol. I. p. 50.)
Trichostomum fontinalioides.
——— fasciculare.
——— canescens. (vol. I. p. 83.)
——— *ellipticum.* (vol. I. p. 58 and 154.)
Syntrichia ruralis.
——— subulata.
Tortula *tortuosa.*

Tortula convoluta.
Catharinea *hercynica*.
——— glabrata. (vol. I. p. 24.)
Polytrichum commune.
——— *alpinum*.
——— *sexangulare*. (vol. I. p. 265.)
——— urnigerum.
——— aloides.
——— *subrotundum*.
Orthotrichum striatum.
Neckera curtipendula.
Bryum androgynum.
——— argenteum
——— *Zierii*.
——— cæspititium.
——— *dealbatum*. (vol. I. p. 58.)
——— hornum.
——— crudum.
——— *turbinatum*.
——— serpyllifolium.
——— pyriforme.
——— dendroides.
Hypnum sericeum.
——— abietinum.
——— *filamentosum*. (vol. I. p. 58.)
——— prælongum.
——— velutinum.

Hypnum *proliferum*.
——— nitens.
——— illecebrum.
——— purum.
——— filicinum.
——— aduncum.
——— *uncinatum*.
——— *revolvens*.
——— denticulatum.
——— triquetrum.
——— squarrosum.
——— cuspidatum.
——— Crista castrensis.
——— cupressiforme.
——— scorpioides.
——— *silesianum*. (vol. I. p. 116.)
Bartramia fontana.
——— *ithyphylla*.
——— pomiformis.
Fontinalis antipyretica.
——— *squamosa*. (vol. I. p. 209 and 260.)
——— *falcata*. (vol. I. p. 260.)
Funaria hygrometrica.
Buxbaumia *foliosa*. (vol. I. p. 88.)

III. HEPATICÆ.

Junger annia *concinnata*.

Jungermannia *julacea*.
——— *asplenioides*. (vol. I. p. 152.)
——— *scalaris*.
——— *Sphagni*.
——— *angulosa*. (vol. I. p. 50 and 161.)
——— *byssoides*.
——— *bicuspidata*.
——— disticha. *Mohr*.
——— albicans.
——— nemorosa.
——— resupinata.
——— complanata.
——— dilatata.
——— ciliaris.
——— epiphylla.
——— pinguis.
——— furcata.
Marchantia polymorpha.
——— hemisphærica.
——— tenella.
Targionia hypophylla.
Blasia pusilla.
Riccia crystallina.
——— glauca. (vol. I. p. 153.)
Anthoceros punctatus. (vol. I. p. 315.)

IV. LICHENES.

Lepraria botryoides.

——— Jolithos.

Lecidea sanguinaria.

——— fusco-atra.

——— fusco-lutea.—About Reikevig. (vol. I. p. 23.)

——— atro-virens. α and γ.

——— pustulata.

Gyrophora glabra. β.

——— deusta.

——— erosa.

——— cylindrica.—Used, in times of scarcity, as food, but more frequently for dying woollen of a brownish-green color.

——— hirsuta.—" Longe optimum in re cibariâ Lichenis genus.—Pagina inferior pilosa. Crescit unicè in lapidibus magnis discretis, et rupibus alpinis, imprimis summis cacuminibus, ubi Falcones sæpiùs insident." *Paulsen in Epist.*

Endocarpon Hedwigii.

Endocarpon *tephroides*.—About Reikevig. (vol. I. p. 23.)

Sphærophoron compressum.

Isidium defraudans.

Urceolaria calcarea.

Parmelia tartarea.

——— subfusca.

——— pallescens.

——— candelaria.

——— *brunnea*.—About Reikevig. (vol. I. p. 23.)

——— gelida.

——— stygia.

——— fahlunensis.

——— omphalodes.

——— saxatilis.

——— stellaris.

——— parietina.

——— olivacea.

——— *scrobiculata*. (vol. I. p. 23.)

——— nigrescens.

——— physodes.

——— furfuracea.

——— ciliaris.

——— prunastri.

——— fraxinea.

Parmelia farinacea.
——— ochroleuca.
——— *sarmentosa.* (vol. I. p. 251.)
——— jubata.
Peltidea venosa.
——— resupinata.
——— canina.
——— aphthosa.
——— crocea.
——— saccata.
Cetraria islandica.—For the account of this and the following species, see Journal, p. 130, and the note at p. 214.
——— nivalis.
Cornicularia lanata.
——— pubescens.
Usnea hirta.
Stereocaulon paschale.
——— *globiferum.*—About Reikevig and other places, not uncommon. (vol. I. p. 23.)
Bæomyces cocciferus.
——— digitatus.
——— deformis.
——— pyxidatus.

Bæomyces cornutus.
——— gracilis.
——— *endivifolius.*—About Reikevig. (vol. I. p. 23.)
——— uncialis.
——— subulatus.
——— rangiferinus.
——— *vermicularis.*
——— *tauricus.*

V. ALGÆ AQUATICÆ.

Fucus serratus.—This, and various other large species of Fucus, serve occasionally for food for the cattle and fuel for the poor natives.
—— vesiculosus.
—— ——— *var.* divaricatus.
—— ——— — excisus.
—— ——— — inflatus.
—— ——— — spiralis.
—— ceranoides.
—— canaliculatus.
—— distichus. *(Fl. Dan.* 351.*)*
—— nodosus.
—— siliquosus.
—— loreus.

Fucus aculeatus.
—— *purpurascens.*
—— lycopodioides.
—— ramentaceus. (vol. I. p. 36.)
—— muscoides.
—— Filum.
—— lanosus. *Mohr.*
—— fastigiatus. *(Fl. Dan.* 351.*)*
—— digitatus.
—— palmatus.—This, the *Sol* of the Icelanders, is the most frequently prepared and eaten of any of the genus. See vol. I. p. 44.
—— esculentus.
—— saccharinus.
—— edulis.
—— sanguineus.
—— ciliatus.
—— crispus.
—— alatus.
—— dentatus.
—— rubens.
—— plumosus.
—— cartilagineus.
—— spermophorus.
—— gigartinus.
—— confervoides.

Fucus *flagelliformis*. (vol. I. p. 36.)
—— plicatus.
—— albus. (*Fl. Dan.* 408.)
—— corneus.
—— fungularis. (*Fl. Dan.* 420.)
Fucus clavatus. *Mohr*.
—— coronopifolius.
—— *fœniculaceus*. (*Conferva. Huds.*) (vol. I. p. 36.)
Tremella lichenoides.
—— verrucosa.
—— hemispherica.
—— adnata.
—— Nostoc.
Ulva umbilicalis.
—— intestinalis.
—— latissima.
—— compressa.
—— pruniformis.
—— lactuca.
—— lanceolata.
—— linza.
—— plicata. *Mohr*.
—— crispa. *auct*.
Rivularia cylindrica. *Wahl. MSS.* (vol. I. p. 86, 100, and 331.)
—— *angulosa*. (vol. I. p. 260.)

Conferva dichotoma.
——— *spiralis.* (vol. I. p. 49.)
——— *bipunctata.*
——— *nitida.*
——— *flavescens.* (vol. I. p. 161.)
——— æruginosa.
——— *vaginata.* (vol. I. p. 50.)
——— *limosa.* (vol. I. p. 160.)
——— littoralis.
——— scoparia.
——— cancellata.
——— polymorpha.
——— rupestris.
——— ægagropila.
——— corallina.
Byssus Cryptarum.

VI. FUNGI.

Agaricus campanulatus.
——— fimetarius.
——— campestris.
Boletus luteus.
——— bovinus.
Helvella atra. (*Fl. Dan.* 354.)
——— æruginosa. (*Fl. Dan.* 354.)
Peziza lentifera.
—— scutellata.

Peziza cupularis.
—— zonalis.
Clavaria coralloides.
—— muscoides.
Lycoperdon Bovista.
Mucor Mucedo.

END OF APPENDIX. E.

APPENDIX. F.

DANISH ORDINANCES

CONCERNING

THE TRADE OF ICELAND

BY

LAND AND SEA;

AS ALSO

THE PRODUCTS OF ITS MANUFACTORIES.

APPENDIX. F.

TRADE OF ICELAND.

§ I.

In what manner the fishing in boats shall be continued by the inhabitants.

THE fishing in boats, being the chief employ of the inhabitants on the coast, is permitted to every person having a fixed residence in Iceland, from which neither the clergy, nor civil officers are excluded. Our appointed magistrates, and especially the inspectors of the districts, have it likewise in charge, that whatever we have ordained, as well as whatever else we may in due time think proper to ordain, be attended to in the strictest manner possible, and also that they shall by all means encourage the inhabitants of the fishing villages to cleanliness, and industry. They have in particular to attend to supplying the fishing villages with good and wholesome water, wherever it may be deficient; and for such purpose the tax-gatherers shall oblige all the fishermen in their several districts, to do the needful work, on pain of punishment according to the nature of the offence. It is our farther will and pleasure, that the lieutenant of the county, the high bailiffs, and bailiffs of theseveral districts, shall give information to our treasury, of whatsoever in their judgment may contribute best and most effectually to the encouragement of the fisheries in general, either by the introduction of new and more suitable regulations, or by removing such obstacles, as may possibly hitherto have laiu in the way; and we shall more especially expect to receive their sentiments, concerning the mode by which the so denominated *loan of men*, may in time be set aside, or restricted; in like manner as it is our intention of doing, on the estates to us belonging.

§ II.

The fishermen are otherwise not to be obstructed in seeking the best places for fishing, and even in their own boats, on condition of their paying to the farmer, on whose possessions they may have taken their stations, for landing their boats, house-room, and services for the season, according to what has been the usual custom at the place. But, in case of all the room being so completely occupied that they cannot find sufficient convenience for themselves, either of houses, or place of landing thereto attached, and that they can find some other landing-place, it shall be permitted them to put into it, and, if they are so inclined, there remain under the cover of tents, or of their own boats. In such case, no claim for ground-rent shall take place, provided the strangers shall not, by treading, have caused any damage in the fields or meadows. In case that they, from deficiency of fish-huts, proper heath, or other places, should be under the necessity of drying their fish near the shore, and in places where grass might possibly grow, it shall not be considered that the farmer thereby sustains any damage, but rather that he therefrom derives a benefit, as the roes, with the fish which are there laid, manure the land, and render it capable of producing grass, if the soil be capable of it; neither can it be deemed hurtful, that fish are dried, even on good grazing-fields, in such parts where the *Vortiden* ends on the twelfth of May. But, in case of any dispute arising between the farmer and the strangers, the bailiff of the place shall nominate two independent persons to survey the place, and estimate the damage, according whereto the high-bailiff will give his decision. But, on the contrary, should the summer-quarter be taken for this purpose, the drying of fish on useful grass meadows cannot be permitted, unless that the consent of the land-holder be thereto previously obtained.

§ III.

In what manner the whale fishery particularly is to be conducted.

In case of any whale being driven on another person's ground, the method prescribed by the rescript of the twenty-third June, 1779, is to be adhered to; and it is farther to be observed, that such of the inhabitants of the country, or strangers, as intend to pursue this fishery, may expect to receive particular assistance for the purpose, according to circumstances, and the utility of the plan. Such persons, more especially, as inhabit the vicinity of the fiords or bays most convenient for such pursuits, as the market-town at Isefiord, and divers other fiords, from Patrix-fiord to the Jökul-fiords, in the west country, and Hvalfiord in the district of Kiose, may expect having assistance granted them for procuring the necessary implements, on due request being made. And, for the same purpose and intent, we will also in such cases give the needful directions for having some young and active persons from those places, duly instructed in the art of the said fishery, and in the manner of using the fish, free of any expence to themselves; and they shall likewise afterwards receive farther assistance for enabling them to establish themselves in Iceland, at some place or other most convenient for such fishery, and there practice the knowledge they have obtained of it.

§ IV.

Encouragement for catching of porpoises, seals, and salmon; as also for herring and flounder fisheries.

It appearing to us as a matter of essential consequence, as well for the inhabitants of Iceland in particular, as also for our dominions in general, that, exclusive of the cod-fishery, now the most pursued, the taking of all other kinds of fish which nature produces near Iceland should be followed; in order to catch as great quantities as possible of all sorts of fish, and that the fish should be cured in such manner that it may not only be used for the support of the inhabitants, but likewise be disposeable as an article of commerce, we would have it taken into consideration in what manner the inhabitants of the country, in case of their exerting themselves in extending and improving the fisheries, more especially with respect to porpoises, seals, and salmon, and also to catching of herrings and flounders, may best be supplied and assisted with the necessary and proper implements; all according to circumstances, and to the statement expected from the magistracy; more especially when such institutions are made on a great scale, and may be productive of any considerable influence in commercial transactions.

§ V.

General rules for the quality of the fish.

The regulations concerning certain descriptions of fish, which existed during the time of the company's charter, can in future, and with regard to particular circumstances, be no longer applicable; but, in case of other, or more sorts of such goods being required by the traders, it must be laid on the basis of this ordnance, in its second chapter, § 4, but nevertheless, and until other regulations shall be fixed among the concerned, those which have hitherto been in use, shall hereafter continue to be observed, and this for so much the greater reason as they are intended to specify the quality of the chief exports from Iceland. Fresh, or soft fish, such as may be considered good merchandize, should be delivered immediately after being taken out of the sea, and untainted; nor must there be any lean or skin-fish among it. The heads must be cut off, the entrails taken out, the fish properly split, in such a manner that the bone be taken out three joints below the navel, and the scales of the cole-fish must be scraped off. Such fish as can be used for dried fish, must be salted immediately on being caught, with the necessary quantity of French salt, or some other sort equally useful. It should be well cleansed, and afterwards properly cured, according to the

Newfoundland mode, in such a manner that it may obtain the proper appearance, and keep well. The neck, and every thing about the neck, must likewise be cut away, before it receives the last day's drying. The dried fish must be well worked and thoroughly dried, and not mouldy, rotten, slimy, or maggotty. The neck must be cut off when it is half dried, or at least before it is received and weighed. The fresh cod-roe must be delivered immediately on its being taken out of the fish, the breeches must be whole, and the roe of a red color, firm, and not spawning. The oil must be clear and clean, and leave no sediment.

§ VI.

Rules for land products, and their most profitable use.

And likewise all such goods as are produced by farming and its different branches, which are sold by the Icelanders, must in general be clean, warrantable, and well worked; whereas, with regard to particulars, the directions contained in the second chapter, § 14, must be attended to. But, as such goods may partly be employed to much better advantage in trade than could hitherto be done, under the influence of a chartered company, such of the inhabitants of the country, as are willing to apply themselves to better modes of cure, may expect to receive some or other suitable encouragement, according to circumstances; and more especially if the plans are of such extent, that they may produce a visible effect in the trade. The same will take place with respect to farming and agriculture, both in general, and also in particular, in so far as any one may put himself forward by considerable or important experiments in gardening, planting of woods, growth of corn, potatoes, and other herbs and roots, or by the proper cultivation of several kinds of Icelandic herbs, useful for food; all according to circumstances, and statement thereon to be made by the magistracy to our treasury, and also partly according to our resolutions hitherto passed on such matters.

§ VII.

Ordinance for the preservation of the eider-duck.

Concerning the catching of birds, and the use to be made thereof, according to law, the same usage shall continue as hitherto has been the invariable custom. But no person whosoever, whether a stranger or an inhabitant of the country, shall be permitted to shoot eider-fowls, or destroy them with dogs or nets, under the penalty of three marks for every bird; and also every one that has been present at the time, and has not immediately reported the offence, shall forfeit half that sum: of which penalty the informer shall receive two-thirds, and the poor of the parish the other one-third, after deducting the expences attending examining into the same, unless it be done on such person's own ground, and in such a manner that the neighbors cannot thereby suffer any loss or damage. But, on the contrary, should this be done on another ground-owner's land, who may have adopted any peculiar method for the preservation of the eider-ducks, either by making islets in the fresh-water lakes, or by building of nests for the reception of such birds, or any other thing of the like nature, he shall pay the damage according to law: and, whoever shall be found guilty of having malevolently destroyed or injured such receptacles, shall be punished by a criminal fine. Neither shall the eggs be taken out of the eyry; and, in such places, where there are several joint owners (unless with the mutual consent of all concerned, and upon certain days for such purpose appointed, and this not to be done at any time later than to the end of the week preceding the week of St. John, at Midsummer), should it be observed that the eider-duck begins to retire to some islet, or other place, which is not the particular property of any individual, there to deposit its eggs, it shall at first be permitted to keep its eggs, both for the purpose of decoying it there and suffering it to increase, until its increase shall have become so visible that the eggs may be taken without detriment to the brood, and in case any of the neighbors or other persons act contrary hereunto, he shall pay one rix-dollar as a fine, of which the informer shall receive two-thirds, and the poor of the parish the other one-third, after a deduction of the law charges; whereas, on the contrary, those who strictly follow such precautions shall be entitled to the rewards stated in our resolution, bearing date the twenty-second of June, 1785.

§ VIII.

Concerning the mode of procedure with sulphur and salt works.

For the farther encouragement and extension of some particular branches of industry, it is likewise our most gracious intention to offer the sulphur-works at Huusevig, and the salt-works at Reykenes, to private adventurers, and on reasonable conditions; for which purpose, those concerned have to apply to our committee for regulating Icelandic affairs. And, should any one be inclined to renew the working of the sulphur-mines, which were formerly in drift near Kreisevig, in the district of Guldbringue, or to work any other considerable mines, or to make farther improvements in the boiling of salt from the sea-water, by the assistance of the hot-springs, such as are to be found in the district of Bardestrand, at Reykholt, and possibly in sundry other places, we would not only herewith cause the same to be permitted, but likewise encourage such endeavors by proportionate bounties. With regard to the sulphur in general, it shall be permitted for every person to make the best use of it, to the extent of his ability, and wheresoever he may find it; with the sole reservation, that this must be done with the consent of the land-owner, in case it should be discovered in any place beyond the limits of the discoverer's own grounds.

§ IX.

Encouragement for using sundry mineral substances.

Of such minerals as are found in this country, whether of earth, sand, or stone, every one is allowed to make free use, provided that they are found on the mountains, public roads, or other places, not the property of any particular person. And, as several of the stony kind are found to be very useful for buildings and other uses, such persons as may be willing to break them, may (on a request for such purpose having been legally made) have a right of property granted them to such places, provided they are not already the property of others; but, should the case be otherwise, they must make an agreement for such purpose with the proprietor of the soil; and the magistracy shall likewise take care that none, without sufficient cause, deny another person the liberty of using such kinds of products, in return for a reasonable acknowledgement, if they are of any consequence, and if he either cannot, or will not, make use of them himself. The same shall be the case with sundry other more rare kinds of mineral substances, such as gypsum and lime-stone, crystal, opal, calcedony, agate, jasper, zeolite, and various sorts of volcanic matter; and even those persons, who may be willing to polish such stones for ornamental purposes (for which experience has proved them to be very suitable), may expect to be recompensed according to circumstances; provided that such an experiment may appear to promise a beneficial result.

§ X.

Encouragement for using sundry other products, and especially of drift-timber.

In like manner, such persons may expect rewards and support, according to circumstances, who shall discover and work pit-coal, or shall find out easier methods of breaking and using surtur-brand from the rocks: for which purpose the magistracy shall, in the manner before described, have the power of making the necessary agreement with the owner of the soil, in the best possible manner. The drift-timber, which is principally found in the northern, north-western, and north-eastern coasts, must doubtless, as heretofore, remain the property of those, who are lords of the shore; but, it having been proved by experience, that the inhabitants, from deficiency of means, cannot sufficiently take advantage thereof, it is our gracious will and pleasure, that, by the interference of the magistracy (at such places, and where it is usual for such kind of timber to be found) in all cases where disputes may arise, an equitable agreement shall be made between the proprietors of the soil, and such others as might possibly intend making a continual use of the place, for their own benefit, with regard to making the best use of such timber. And, under such circumstances, the undertakers of all such concerns may expect having some or other gratuity allowed them, proportionable to the purpose intended; and more especially in so far as they may intend to make use of the said timber for building of large boats, and other vessels fit for fishing by sea, and the carriage of goods; for which purpose, those places where such advantages happen, stand in greater need than others along the coast.

§ XI.

Of what advantages will be granted to such tradespeople as are most useful to the country.

Tradespeople, who may be inclined to settle in Iceland, may do so without hindrance or molestation, not only in the trading towns, but even in the country, it being permitted, by the placard of the eighteenth of August, 1786, in its 14th article, for every one to carry on whatever lawful trade he may think proper in the country, by the best means in his power, mercantile business only excepted, as expressed in § 2, of the second chapter of this ordinance. Such tradesmen as are of the greatest utility to the country may likewise expect to have a royal assistance granted them for their establishing themselves in some or other of the trading towns in Iceland; such as weavers, hat-makers, fellmongers, ropemakers, blacksmiths and whitesmiths, joiners, coopers, house and ship carpenters, bricklayers, and stone-cutters. Such people whether they are masters, or journeymen who have perfectly learnt their trade, may in consequence apply to our treasury, where they will receive the necessary information of what advantages may be granted them, according to circumstances, and the several local situations: but before that they come forward with such requests, they must be duly provided with attests from the magistracy whom it may concern, and also with the needful proofs of their capacity. Such, and, according to circumstances, still greater advantages will be granted to those, who may establish larger manufactories, and those that may prove more useful to the country (some of which, such as ropemaking and furriery, have already in former times been followed with good success), and, more especially, if occasion should thence arise of using the raw products of the country in a more beneficial manner, or if the wants of the country might thereby be the more easily supplied; under which head come fish, glue, and isinglass: the making use of horse-hair for various purposes, to which it may be rendered serviceable; different sorts of earth and clay for painting colors, and several other means of support, which nature appears to have thrown in the way.

§ XII.

In what manner the manufacture of woollen goods, hitherto used in the country, shall be continued.

Woollen goods having, from the most ancient times, been the chief manufacture of the Icelanders, and having in general found a tolerably good demand, partly even in foreign countries, their own profit will probably stimulate the inhabitants in future to continue exerting their utmost industry in the improvement of this kind of goods. But what different sorts thereof would now be most profitable for them to work, and of what quality they ought to be, will in future solely depend upon a mutual agreement between the traders and the inhabitants; and, consequently, the traders themselves must furnish the inhabitants with such samples and information, according to which they may deem it most advantageous for the latter to work. Now, when the inhabitants have undertaken, according to such samples, information, or patterns, to make the goods bespoken, whether they be mittens, stockings, or woollen stuffs, they ought diligently to endeavor to follow such directions; it will consequently in the present case, solely depend upon such goods being duly and properly manufactured, in regard to which the directions contained in § 14 of the second chapter are to be strictly followed; whereas, on the contrary, all the directions and methods which have hitherto been recommended by our colleges and our magisterial persons, relative to the manufacture, &c., of sundry kinds of woollen goods, heretofore in use, are herewith annulled and put out of force, from the first January, 1788. But as the ground on which the foresaid ordinances are invalidated, is solely this, that they refer to sundry specified kinds of goods, which may undergo much alteration by opening a free trade; so on the other hand, the inhabitants of the country, in as much as they manufacture the commodities hitherto in use, must continue so to do in such manner as has been heretofore prescribed. The magistracy have therefore, when any disputes arise, as well in these, as in other cases, carefully to examine and determine, whether the articles actually are good and sufficient or not, although that the particular regulations hitherto prescribed may no longer be applicable.

§ XIII.

Encouragements for improving the fabrication of woollen goods.

We having so much the greater reason to expect that the manufacture of woollen goods will in future be improved, so that the inhabitants of Iceland may hereafter fabricate different kinds of woollen stuffs, applicable to different uses, and which may be brought to greater perfection, whenever a better mode of spinning and weaving shall become more general among them; we should learn, with the most gracious satisfaction, that any of our own subjects or foreigners would use their endeavors to introduce some or other kind of beneficial manufactory, by which the great quantity of wool grown there might be used with profit for all our other states and dominions; in which case, the undertakers of such establishments may expect having suitable encouragement given them for providing against the first attendant expences, or for promoting the sale of the articles manufactured. The same shall also take place with regard to the manufactory now established at Reikevig (which it is our most gracious intent to yield upon very reasonable conditions, to such persons who may therefore make application to our commissioners for the managing of the Icelandic commerce), as well as to several other institutions of the like nature; such as providing the members of the workhouse with work that is most suitable for them, raising a dyer's house and a stamping-mill in the north country, and whatever other improvements may be made in the woollen manufactory.

§ XIV.

Concerning the encouragements.

Moreover such weavers of woollen cloth and linen who are inclined to establish themselves, in some trading town in Iceland, may expect that suitable encouragement will be granted them, according to circumstances, for their establishment, and for the prosecution of their business; and we therefore do, not only for the present, promise every weaver of woollen or linen a proportionable premium for every person whom they shall prove to have been duly taught the art of spinning and weaving, but also that, if any of those persons, who (according to the plan by us laid out in the year 1780, for our kingdom of Denmark) have been taught in the manufactories established in these kingdoms, shall be disposed to establish themselves in Iceland, for the purpose of their following the profession that they have learned; the first twelve that may so offer themselves, shall, besides, be entitled to a premium of twenty rix-dollars annually for the first twenty years; but in such case it will be required, that they are provided with a certificate, duly attested by the master of the factory, of their being thoroughly capacitated, and also declaring that they shall continue their profession for that space of time, and continually keep one boy in their employ, who shall be fully instructed in their business in the course of five years. And besides which, a magistrate, or some other civil officer, shall, together with a clergyman, be able to attest that he has actually and continually been employed in the profession taught him, and that his master has fulfilled his obligation of having at least one apprentice, who shall have been fully intructed in the course of five years. And every person, either woollen or linen weavers, who shall demand a premium for having brought up apprentices to his business, must produce an attest, as beforesaid, from some magisterial person, proving that the object is thoroughly instructed in his art, and likewise produce in court a sample, given in to the proper magistrate, of the ability of the apprentice. In case of any foreign master tradesman, or journeyman, who has thoroughly learned his business, applying to the magistracy in any of our sea-port towns, for a passage over to Iceland, with intent there to settle, he may immediately, on giving sufficient proofs of his ability, expect likewise to enjoy all the advantages held out, and also to participate in the freedoms allowed to foreigners in general in Iceland, by the placard of the eighteenth August, and the ordinance of the seventeenth November of last year.

§ XV.

Premiums for spinning and weaving of woollens and linens.

It is besides our will and pleasure, that (for the better encouragement of the inhabitants to attend, with all possible industry, to such manufactory, a peculiar premium be appointed for woollen goods made in Iceland, from well-assorted wool, and of the finest quality; and also previously, for the first ten years, a premium of £10 per cent. on the value of the cotton, flax, and hemp-yarn there spun by tale-reels. It is, however, incumbent on those who apply for such premiums for yarns, to produce the needful custom-house attests on the quantities so imported from Iceland into our kingdoms and provinces, as likewise to make attestation on oath, both of the value, and also of its being actually spun in Iceland; on which point, however, we will not, for the said space of ten years, prescribe any fixed rule, neither with regard to the fineness of the spinning, but leave the same entirely to the agreement made between the parties concerned, it being their joint interest to have the yarn spun to the utmost fineness that the quality of the wool will permit. It is likewise our will that a suitable premium be allowed, for the same space of time, for weaving of coarse cloths and linens, chiefly used in Iceland, in such wise, that a premium of half a skilling per ell shall be paid for every ell of coarse cloths and linens; and one skilling for every ell of finer cloths or linens, whilst still in the loom; the persons concerned providing themselves with an attest from a magisterial person, and a clergyman, of the size and quality of the piece. In like manner, such persons as have particularly exerted themselves in spinning and weaving, may expect, that they, in case of their being deficient of the necessary implements, will be gratuitously supplied by us, in future, as hitherto, with some Danish looms, spinning-wheels, hasps, and cards, all according to circumstances, and proposals to be made by the magistrates.

§ XVI.

Institutions for teaching capable pupils in spinning and weaving of woollens and linens.

It will likewise depend upon circumstances, in how far the institution which we have set on foot for causing young, and willing Icelanders to be taught in the art of spinning and weaving of woollen and linen goods (partly at the manufacture of Reikevig, in Iceland, and partly in other manufactories, here in Denmark), on condition of their again returning to Iceland, and there practising the knowledge that has been taught them, will be continued. The foresaid institution at Reikevig, shall be continued until the term of years appointed is expired; or until the fixed number of fifteen pupils, three from each of the five districts of Oefiord, Hunevand, Rangevalle, Myhre, and Arnæs, shall be thoroughly instructed, both in spinning and weaving. During the time that they are so learning, they shall have free board and lodging, and the needful clothing for daily use: and, although that the term of apprenticeship is generally fixed at three or four years, yet this shall not be any impediment to those who may within a shorter time, so perfect themselves, that they may be deemed fit to work for themselves, and afterwards to teach others, immediately returning home again. The district judges must, however, not at present nor hereafter, propose other subjects for being taught at our expence, than such, of whom they may with certainty expect, that, after having been completely taught, will again return and carry on their profession; and if any one should request to have his children or relations taken to learn the business, he shall also oblige himself, after their being properly educated for it, to provide them with opportunity for carrying on their work. It being intended that such youths, as soon as they have obtained the needful proficiency, should again return, and settle in the districts from whence they were taken, in order there to extend to others what they have themselves been taught. For which purpose, each man shall not only be provided with a loom, and every female with one or two good spinning-wheels, but they shall also receive some assistance towards their establishment, and afterwards receive two rix-dollars premium for every one, whom they shall have educated in that country, during the first twenty years; and to the persons taught, will likewise be given some implements for their business gratis, all on condition of the same being duly proved, according to § 14 of this chapter.

According wherto, we herewith direct and command our counts, commanders of districts, barons, bishops, judges, presidents, burgomasters and councils, sheriffs, magistrates, and all others, to whom these presents may come, that they cause the same to be immediately read and published in all proper places, for general information. Given at our palace of Christiansborg, in our royal residential city of Copenhagen, the thirteenth of June, 1787.

Given under our royal hand and seal.

CHRISTIAN R.

Seal.

REVENTLOW. HANSEN.
COLBIORNSEN. VORNDRAN.

PLACARD,

Whereby sundry articles, concerning the trade to Iceland, are more specifically laid down.

We, his royal majesty's, the king of Denmark and Norway, &c., &c., appointed president, deputies, and assessors in the chamber of taxes and interest, do herewith certify, that, it having appeared that the mercantile persons in Iceland have, in some wise, misconstrued some parts, both in the ordinance of the thirteenth of June, 1787, and also in other regulations issued concerning the trade and navigation to Iceland, in a quite contrary manner to the meaning and intent of the said ordinances, it has pleased his majesty, by his resolution, bearing date the thirtieth of last month, to deem it needful to give a further elucidation to the said parts or passages, according whereto the magistracy of the districts therein concerned have to act and direct themselves, so as to prevent all fraud and deceit on behalf or part of the traders. And therefore his majesty, for the purpose of the intent of the said most gracious resolution being the better carried into effect, has graciously been pleased to add the following articles, to be observed and obeyed by all whom it may concern. These parts are, the placard concerning throwing open the Iceland trade, bearing date the eighteenth of August, 1786, in its 13th §; the ordinances concerning the liberties granted to the trading towns, now forming in Iceland, bearing date the seventeenth of November, 1786, in its 15th §; and the before-mentioned ordinance of the thirteenth of June, 1787, in its second chapter, §§ 2 and 11; several traders having taken advantage thereof, to extend their trade to a far greater degree than is expressed or specified in the ordinances given concerning the free trade, when taken in their proper intent and meaning. In order therefore to prevent such abuses, which, in process of time, might be productive of the most dangerous consequences to the commercial trade in general, his royal majesty has been most graciously pleased to ordain, that the privileges, which are granted to the trading towns, by the ordinance of the seventeenth of November, 1786, in its 4th, 6th, and

7th §§, compared with the placard bearing date the eighteenth August, next preceding, in its 12th §, on which the parts before-stated are founded (namely the 13th § of the placard; the ordinance concerning the privileges of the trading towns, in its 15th §; and the ordinance of the thirteenth of June, in its second chapter, 2nd and 11th §§), do not refer to others than such as keep bed and board in the trading towns, and carry on a constant trade there; and likewise, that the country traders, agreeably to the ordinance of the thirteenth of June, 1787, in its second chapter, 12th and 13th §§, shall likewise be under the obligation of carrying on the trade, as well in winter as in summer, the result whereof is:—

See the placard of date twenty-third of April, 1793, § 1.

(*a*) That what is contained in the placard 15th §, the ordinance concerning the privileges of the trading towns, in its 15th §, and the second chapter, 2nd and 11th §§, concerning such citizens, is not to be otherwise understood, than regarding such as have established themselves in the trading towns, and as carry on trade there both in winter and summer, either by themselves, or by their acknowledged factor, except what is otherwise mentioned in the last-cited four places, viz. that they may trade to or with whatever foreign place they will, cannot, if meant to be out of the district, be otherwise understood, than as merely applicable to the special trade of the ships, which shall have lain four weeks, without having commenced any trade with the country; either by having thrown up some sheds, or by having there erected some building; and they shall likewise be allowed to lie and carry on trade, but only in the proper trading towns, or in such out-ports as are either permitted by the liberation of the trade, or may hereafter be erected, with the approbation of our chamber of taxes and interest.—

See placard of twenty-third April, 1793, § 1 and 2.

(*b*) That no man, who shall have established himself in any trading town, or in the district belonging to it, and shall have taken his burghership there, either personally, or by his factor, shall establish any trade in any other trading town, or in the district thereunto appertaining, under the plea of having, either by himself, or by his factor taken his burghership at such place; or under the pretext of the last-mentioned carrying on trade there on his own account; and therefore he shall be obliged (in case of the magistracy finding his assertion of carrying on trade there for his own account, to be liable to suspicion), to make deposition upon oath before the court, that all the goods in which he means to trade, as well import as export goods, are to him solely belonging, and not to any person in any other district.

See placard of twenty-third April, 1793, 1, 2, and 3 §§.

(*c*) That such persons, belonging to Denmark, Norway, or the provinces, who are inclined to carry on a trade in the country, shall likewise form an establishment in the trading towns, raise buildings, and there take burghership, or at least so do by their factor, who in such case shall keep his house and office, and carry on trade, both in winter and summer, without doing which, he may not trade with any inland town, excepting as a special trader, for the space of four weeks; and consequently he must not employ any other factor in his stead, and have such a person made a burgher, and still less shall he employ any farmer in carrying on trade for him during the winter.

According whereto, all persons concerned have to govern themselves.

Chamber of Taxes and Interest, 1st June, 1792.

(Signed.)

REVENTLOW. HOE.
HANSEN. COLBIORNSEN.

KOLLE. SCHIONNING. HANSEN. WORMSKIOLD. BUDT. M. VON ESSEN. MANKE. HAMELEFF. JOHANSEN. VORNDRAN. WADRIN.
GUNDELACH.
SULDEN.

23rd April, 1793.

PLACARD,

Whereby sundry passages in the Placard of the 1st of June, 1792, are more particularly explained, for prevention of the establishment of any prohibited trade in Iceland, and of other misuses in trade, which have there taken place.

We, his royal majesty's, the king of Denmark and Norway, &c., &c., appointed president, deputies, and assessors in the chamber of taxes and interest, do herewith certify, that his royal majesty having, by his resolution of date the thirtieth of May last year, which was made public by the chamber of taxes, &c., in a placard of the first of June next following, been graciously pleased to give a more full explanation concerning some passages, both in the ordinance of the thirteeenth of January, 1787, concerning the commerce and navigation of Iceland, and also in sundry other regulations thereunto appertaining, which have in part been misunderstood, for the prevention of fraud and deceit on the part of the traders; his majesty has now again thought proper, in order that the intent of the said most gracious resolution may the better be accomplished, to give the following articles, bearing date the seventeenth of this month, for the observation and notice of all concerned.

1, That, exclusive of what is ordained in the foresaid placard, letter *a,* it shall be strictly prohibited to expose goods for sale to the farmers, or any others whomsoever, or to send any persons out with goods for sale into the country; and also that whosoever shall be found to have acted hereunto contrary, or against what is ordained in the placard aforesaid, shall be punished by a mulct of one-fourth part of the value of the goods in which such trade is carried on, as well of the cargo brought on shore, as of the Icelandic goods so purchased; and whether it be the master or the mate of a vessel, who has undertaken to carry on such a trade, he shall besides be fined twenty rix-dollars; and such Icelanders, or other persons established in the island, who shall suffer themselves to be detected in such illicit trade, shall likewise incur a penalty of from five to ten rix-dollars, according to circumstances.

2, The same punishment shall likewise be inflicted on the traders established in Iceland, who shall act contrary to what is ordained per letter *b*, in the placard. The traders from Denmark, Norway, and the provinces, who, in their trade to Iceland, shall omit attending to the contents of the placard in letter *c*, shall be punished in like manner.

3, But, on the contrary, if the traders from Denmark, Norway, and the provinces, shall, agreeably to what is directed by the placard, in letter *c*, establish a fixed trade in the trading towns, and there keep house and office, or at least do so by their factor, they shall have a right to trade with any inland towns thereto authorised, in the district of such trading town; and such factors shall, agreeably to the placard, be obliged to take their burghership; whereas such factors and other servants, as are employed by merchants who have taken the houses of trade effects and materials and who have themselves taken their burghership, shall be excused from taking burghership.

4, It shall be totally prohibited to grant letters of burghership to the peasantry; and such letters of burghership, as may have already been granted to such persons, shall be revoked, and annulled, unless they are willing to establish themselves in the trading towns, and there only carry on trade as burghers according to their privileges; but they shall in no wise carry on trade at their farms, under penalty of being mulcted as aforesaid.

5, If any free or private trader shall be found to remain and carry on trade at the trading town, after the expiration of the four weeks granted by agreement; or at the authorized out-ports, and there carry on trade; then, and in such case, all such persons, be they masters, mates, or seamen, shall pay a penalty of four rix-dollars for every day, that it shall be lawfully proved they have staid above the said time.

6, And likewise, such private traders, as live in out-ports, where there has not been any proper or authorized trading town, or in the fiords and bays, for the purpose of trading, shall be subject to the penalties ascertained by the 1st §.

7, One fourth part of the fines aforesaid shall be given to the informer, one fourth part falls to the justiciary box, one fourth to the house of correction, and the other fourth part, if the fines are under ten rix-dollars, to the poor of the parish; but, if it exceeds ten rix-dollars, it shall be given to the poor of the whole district; and the merchant, in whose district such unlawful trade is carried on, shall not only be authorized to give information thereof, but also, in the absence of the magistracy, shall have power, with the assistance of people employed for the purpose, to lay a sequester on the goods, on condition of his immediately requiring the magistracy to take a legal survey and estimate on the same, and thereafter to proceed in the cause to its conclusion, according to agreement, and at the expence of the person implicated.

According whereto, all persons have to conduct themselves.

Chamber of Taxes and Interest, 23rd April, 1792.

(Signed)

REVENTLOW. HOE.
HANSEN. COLBIORNSEN.
KOLLE. SCHIONNING. WORMSKIOLD. BUDT.
M. VON ESSEN. MEINCKE. HAMELEFF.
JOHANSEN. VORNDRAN. WADUM.

GUNDELACH,
Clerk.

FINIS.

INDEX.

INDEX.

INDEX.

INDEX.

INDEX.

INDEX.

INDEX.

Printed in the United States
By Bookmasters